THEORY OF
IDENTITIES

FRANÇOIS LARUELLE

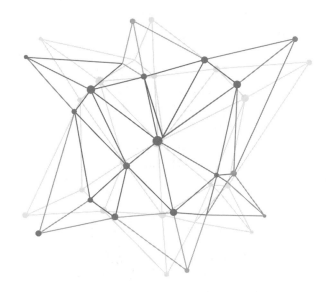

THEORY OF IDENTITIES

TRANSLATED BY ALYOSHA EDLEBI

COLUMBIA UNIVERSITY PRESS

———

NEW YORK

Columbia University Press
Publishers Since 1893
New York   Chichester, West Sussex
cup.columbia.edu

Théorie des identités: Fractalité généralisée et philosophie artificielle
copyright © 1992 Presses Universitaires de France

Copyright © 2016 Columbia University Press
All rights reserved

Cet ouvrage a bénéficié du soutien des programmes d'aide à la
publication de l'Institut Français.
This work, published as part of a program of aid for publication,
received support from the Institut Français.

Library of Congress Cataloging-in-Publication Data
Names: Laruelle, François, author.
Title: Theory of identities / François Laruelle; translated by Alyosha Edlebi.
Other titles: Théorie des identités. English
Description: New York: Columbia University Press, 2016. |
Includes bibliographical references and index.
Identifiers: LCCN 2015039843 | ISBN 9780231168946 (cloth: alk. paper) |
ISBN 9780231541459 (e-book)
Subjects: LCSH: Philosophy. | Science—Philosophy. |
Identity (Philosophical concept)
Classification: LCC B77 .L3713 2016 | DDC 111/.82—dc23
LC record available at http://lccn.loc.gov/2015039843

Columbia University Press books are printed on permanent
and durable acid-free paper.
This book is printed on paper with recycled content.
Printed in the United States of America
c 10 9 8 7 6 5 4 3 2 1

Cover design: Lisa Hamm
Cover image: © Nigel Grima. Private Collection/Bridgeman Images

References to websites (URLs) were accurate at the time of writing.
Neither the author nor Columbia University Press is responsible for URLs
that may have expired or changed since the manuscript was prepared.

ambient with a grant

Figure Foundation

CONTENTS

PART III

Principles of an Artificial Philosophy

PREFACE TO
THE ENGLISH EDITION

RETROSPECTION (2014)

PHILOSOPHERS always present their undertaking with the seriousness of ferrets at work and usually without much humor. Saving their irony for the world or for their predecessors, they are more reluctant to take on their own fictions, their "novels." And yet they dream. Plato and Nietzsche suggest that philosophers "dream" of the fumes of intoxication rather than of intoxication itself, of poetic inspiration rather than of poetry. In most cases, their dreams come very close to reality, lingering for a long time over concepts. But these reveries are philosophy's matrix, its preparation or the lying-in-wait of its rationalization, the way Descartes presents his dreams in order to prepare the exposition of his cogito. Others, wide-awake, still fantasize about chaos, explosion, breakage, rupture, and collision, but also, and quite differently, tsunami or drowning, always like Descartes. These are the philosopher's dreams within the limits of simple reason, according to Kant, that cold dreamer, one of the most humorous perhaps, who was surprised to find himself dreaming of Swedenborg.

As for us, it is time to sound the alarm and to clarify an ambiguous situation. We hesitate between two modes of classification grounded in the style of our object, a corpuscular mode earlier and—given the final impossibility of maintaining this corpuscular distinction between stages or eras—an undulatory mode now. Each of non-philosophy's works contains all of this in an indivisible or "holistic" way. But even the division into stages or steps serves to mark doses or proportions, nuances and accents that should not be dogmatically isolated. Non-philosophy is a system of stages as much as of phases, of particles as much as of flows or waves. It is risky to prefer the classification by steps or stages and to fail to notice the waves that sweep them away and render every classification undecidable.

No doubt we too have dreamed. We dreamed of a joyous Iliad that left too hastily to look for its Odyssey, that believed it reached this Odyssey from time to time, and that wandered from shore to shore without ever finding solid ground, doomed to sink in the contemplation of immobile stars that the river Metaphysics carries along. Let's say immediately and clearly: non-philosophy is a dreamed philosophy, a reverie or a fiction that owes a great deal to a certain power of dreaming peculiar to music.

Theory of Identities is the book of the milieu, of the middle. Such is its situation, the one from which a balance sheet starts to become possible. It presents itself as a balance sheet in a doubly forced writing—forced by the problematic results of the Philosophical Enterprise and by the crucial question: "how can we continue with so weak a productivity?" Let's think for a moment in terms of capitalist humor, i.e., in prosaic terms: philosophers have spent lavishly and imprudently, overindulged the zero unemployment, and consumed without restraint the abundance of materials. Let's have them work "part time" and experience a labor scarcity. Most of them do not worry about their means of subsistence; the decline of these means is not their problem. But what can existence and the whole of ontology do without means, if they are merely consumptive, i.e., if they simply reproduce their own sterility?

The work's *milieu*, its middle, is a way of rejecting the system. And there are many possible rejections. Treated negatively as a caesura, the milieu determines a philosophy's fairly rigid and definite order of exposition (Heidegger and Wittgenstein). This is, for us, the sure sign of a will-to-system and to-unity that failed. But it does not always divide in half a work that will be stitched and disjointed a thousand times by the interpreters. Treated more positively as a continuous milieu of existence, it does not exclude modalities, nuances, and accents. We began by practicing the care of the One against the care of Being, but we have also cut and recut a great deal in the classical concepts. The One has become the One-in-One, then the under-One, then Idempotence. The classical duality has become the unilateral duality or the complementarity. The ekstatic vector was divested of the double ekstasis and has become underekstasis, identity a nonidentity. Nonphilosophy is a thought of the duality that began through a reclamation of the One rather than of Being. But the One founds a very different lineage (One-in-One, flux, duality, under-One, before-priority) from Being's (priority, difference, stases or stages or figures) and leads more reliably to duality than Being does, which aspires to the One and is satisfied with that aspiration. Under this complex form, *Theory of Identities* is—after *Biography of the Ordinary Man*—the retrospection on the construction of a philosophy, the central arc of a bridge open to the circulation between the first construction of non-philosophy and its current or "nonstandard" forms. It is the *mise en tableau*, the visual concentrate of a long caesura that shapes philosophy's intimacy through science's means and that appears here according to a vertical cut, which clearly shows what the battle between these two thoughts (science and philosophy) has been. In this sense it is a synoptic book, a systematic capitalization of acquired knowledge, a book that crosses once more the various possible paths within non-philosophy, as the digraphia of a balance sheet, recording the advances and retreats, at times lacking the balance and symmetry of later works. It brushes against the encyclopedia, without returning to Noah's arch of absolute knowing in which Hegel recollected history.

Let's bracket the dream (even though we will have to come back to it). Is there anything real left in this book and in the entirety of our work? There are three possible strata of interpretation. Superficially, a partition of the still very academic material according to classical themes and, above all, according to stages or eras. More profoundly, a quantum distribution by waves and corpuscles. Lastly, a musical affect, at least an aesthetic inspiration of the whole, which can be detected ever since the first text on Ravaisson.

The most superficial appearance is the exoteric, scholastic, and contingent multiplicity of courses, a tradition that stems from Aristotle, with its transformed remnants in *Theory of Identities* and in the moderns of stratification from Hegel to Badiou, according to objects and domains (ethics, ontology, psychology, logic). Within this, we introduce the central theme of identity as a weapon against the postmodern theme of difference, but it does not yet reach quantum indivisibility. We carve out an essence of science; then an essence of non-philosophy and of its fundamental concepts like determination-in-the-last-instance; then generalized fractality, which is conjugated with philosophy as a terrain for the exercise of this problematic. Here fractality anticipates the future use of quantum mechanics. Generalized fractality rather than textuality: this *mise en terrain* of the problematic corresponds to some of our "experimental" contemporary texts, which exhibit an artistic fractality. Lastly, its application as "artificial philosophy," which anticipates a mode of nonstandard philosophy or of philosophy through the quantum model.

The question of stages is reifying. It engenders continuous appearances and dogmatic and unilateral interpretations. Non-philosophy is an elementary material, but it can be distributed differently through waves and oscillations. The undulatory and particulate form is not just an image or a metaphor. Each work has a near-absolute autonomy; each restarts the trajectory that assembles the various phases or brings the entire problematic into play. This is the corpuscular aspect with its various emphases, which are so many oscillations, so much so that each of non-philosophy's works as well as their totality give the impression of a wave or a thrust and reject as superficial philosophy's classical distinctions. Hence the replacement of the division-by-stages—which appeared inadequate and surpassed—with another more supple and more individualized schema, no doubt more in harmony with the current schema of a quantification of the work (in the "qualitative" sense that quantification takes in physics rather than in the sense of the quantity of books and titles).

Nevertheless, we are still dealing with a surface-inventory that has to move from discursive themes to another model, for example to leitmotifs à la Wagner. Non-philosophy is doubled more globally by a musical organization or tissue. Vertically, it is a spiraled thought, contrapuntal in spirit or with superposed themes (in a musical rather than quantum sense, but the former announces the latter). Horizontally, it is a melody that exposes and reexposes the themes. Its profound or desired model is musical. To be sure, its form is still too classical and insufficiently inventive, and nowadays it dares to go beyond the academic form only in some experimental texts grounded in repetition. It is born of relatively precise obsessions, of repetition through a system of variations, the ideal of a repetitive or variational thought from the great classical models (Bach, Beethoven, Brahms, Wagner) up to the most recent (Cage). A work is a retrospective self-knowledge, the permanence of solutions and of obsessions, a theoretical confession of faith. And music is the placenta that has to give birth to non-philosophy.

What does it mean to philosophize in a nonstandard way? *Theory of Identities* is a moment in the search for a new practice. We can now expound a few of its traits. It always involves philosophy, but in new relations to science, to art, or to religion. It is not a matter of a new "philosophy of . . ." nor, on the other hand, a scientific, artistic, or religious practice of philosophy. We touch on all these old relations, but their confusions would denature our project.

On the side of the knowings [*savoirs*] that are our materials (empirical but also philosophical knowings, which we no longer distinguish as the philosophical tradition does), what we claim as "gnosis" in a modernized sense is a double knowing: it pertains to being and to having. On one hand: the one we "are" for the having that is acquired and sedimented under the rubric of universal humanity and that we all assume; on the other, the one we are currently acquiring the knowledge [*connaissance*] of or the one we "are" only under the rubric of non-philosophy. Obviously we are no longer dealing with traditional relations founded on the hierarchical distinction between philosophy and some knowings, since philosophy now forms part of these empirical knowings. This old distribution is replaced by another, indexed to being and having, to the historical tradition and the current acquisition of a new knowledge via "nonphilosophical" paths or procedures. This is the distinction between universal knowing (which seeks to be total in the sense of philosophy) and the knowing we will dub "generic" "or "man-oriented" instead of "philosophy-oriented."

We are not specialists of the postures called science, art, and religion (we are not even specialists of philosophy, which, to begin with, is any posture whatsoever). But we assume them each time as materials or variable properties of an object = X to be determined. We obviously assume them as our virtually universal being, which participates in this acquisition of the primitive reserve of knowings. But we also assume them more profoundly under the current rubric of non-philosophy, which deals with "having" or acquiring a knowledge of a new type and of another destination, a knowledge that no longer enters into this primitive reserve but makes use of it in order to go beyond it. For this reason, we posit the identity of the variables of science and of philosophy, to be conjugated in view of this object = X. But we also distinguish as quantum and generic the knowledge of this knowing that has a new object of its own, neither that of science (of art, of religion) nor of philosophy. It is the generic object par excellence, which can only be humanity. Not the humanity of humanism, but the humanity of the last-instance.

How can we gain or produce this second knowing, the knowledge we call "gnostic"? The first operation of the method, the first product, is specifically quantum in spirit and consists in reciprocally weakening the brute spontaneity of the two opponents and in transforming them into simple inventive, productive, or variable forces for a nonart, a nonethics . . . that will therefore be deprived of every principial [*principielle*] and dominant value. This first product of variables (multiplications = interpretations) derealizes and idealizes the science that is extracted from the real and interiorized to the concept. At the same time, and inversely, the second product, the inverse of the first, deprives the concept as brute variable of its instinct for domination and does so by means of quantum algebra (the imaginary number). We thus transform the brute properties of the primitive reserve into variables (nonbrute this time, but quantum) within the chamber where the struggle between disciplines unfold. And we prepare a wholly new—and this time generic—determination of the struggle, which will interiorize in its own way the quantum struggle that neutralized the opponents.

The outcome is an equalization of variables and a quantum knowledge, probable rather than certain, a way of asserting quantum indetermination. For the moment, and at this stage, the indetermination is neutral; it holds indifferently for the two sides of the conflict. This method of reciprocal double interpretation is a symmetry, give or take an inversion. The products' inversion implies noncommutativity itself for the symmetrical instant; it implies the primacy of one or the other variable or of any product whatever. Symmetry is still formal or indifferent and, in this sense, has to be broken. We still have to decide which variable will sweep it away, if this role is not assigned to philosophy on its own, science on its own, or to one of their products.

Hence a second operation: a new (this time "generic") decision has to be grafted onto the quantum. It will no longer be a decision of overdetermination or of mutual reinforcement in view of a superior knowing à la philosophers. It will instead be a subtraction by the real, i.e., by science, which ratifies quantum noncommutativity. A third instance, a third term, is needed to materially break or realize this symmetry, which is still entirely formal.

Either the quantum symmetry remains in itself as a virtual third term or instance of overhanging [*surplomb*] or the break in symmetry pertains to a nonquantum origin, even if it also affects the quantum. It is a matter of actualizing the quantum and its still-formal indetermination. We cannot stop at this pure formalism of products. We have to tie them together, if only to find a real order between them; this order makes possible a nonformal inversion that does not remain external to the hands of an agent.

The instance that has to carry out this passage from the quantum to the generic and break the quantum's still-formal symmetry—by realizing it concretely in the experience of the real—is what we call idempotence, a fundamental algebraic notion. *Idempotence is what subsists of the One when it is affected by the imaginary number.* In reality, we distinguish the double transcendence or the doublet of the philosophical variable and the absolute instance of the One-of-One, which is also specifically affected by the imaginary number of the quantum real. We introduce here the One that finishes the philosophical edifice. And so, there are three variables to be considered and not only two. The One must also be brought face to face with the imaginary number, it has to be affected by it as idempotence. But this number, at least insofar as it is multiplied or interpreted by philosophy as a simple variable, is also stripped of its philosophical sense in order to become a variable or a productive force, as we have already said. We understand by this that the productive forces are multiplied by one another and are so retroactively from the generic One of idempotence, an instance that is superior to transcendence itself. In one way or another, and in the more or less long term, all the strata of philosophy will be affected and reduced by the imaginary number or the real. The paradox is that the formalism, which continues to reside in the quantum, is realized and broken only if an instance intervenes that breaks the formal symmetry by relating it to human experience. And this instance has to be identical to itself, but also capable of tying together the three instances, which are all affected and reduced by the imaginary number. This general reduction of philosophy, i.e., of the three stages that constitute it as real (in the narrow sense of the quantum real), as reality (in the sense of the now-simplified transcendence), and lastly as absolute (become strictly radical), which moves from the theory or image of the world to the theory or image of humans, constitutes it into a correlate of the generic humanity of the last-instance. Non-philosophy is *also* the unification of quantum theory and of Marxism.

We have surpassed the thematic and even the theoretical level of *Theory of Identities*. This suggests the direction in which our future research will develop. Our practice mobilized scientific as well as other models (quantum, Gödelism, Euclidism, photography, theology . . .). These are forms of dual thought; they conjugate two objects or two postures. The one is not specifically philosophical, but the other is necessarily so, as if it were each time a matter of the unification of philosophy (which has become a simple empirical theory) with an empirical discipline elevated to the state of paradigms. Nonphilosophical practice has no proper and well carved-out object; it seeks fluid models of the undulatory, the fluvial, and the oceanic rather than the topological. Or again: the river's unhurried drift, the casualness of the great wave, the furor of the tsunami, and among them the hesitation of the drunken boat that transports us. These models are purified of their positivity, universalized and idealized. They become Ideas or paradigms. Scientific, artistic, and religious theories in their multiplicity are our Ideas, our paradigms. This is not a dogmatic and terrorist purification of a discipline, which is transformed into a constraining model, but a double transformation, a purifying idealization, a realizing subtraction, and, in the middle, an idempotence of the drunken boat.

Some philosophers have almost understood that musical metaphysics merits the name "meta-physics" only if it is a microphysics. Provided we understand it as a quantum superposition and not as an identification, nonphilosophy (at least the one of which we philosophers pursue the dream, failing to perform it—but does it need to be performed a second time?) is theory interpreted as music and music interpreted as theory. The musical form is not so obvious. It has no doubt been dreamed more than executed and so has the theoretical form. But this is the most profound, perhaps the most absurd dream. A multiplication of music and theory by each other as variables, their noncommutativity, their fusion as generic human properties, the utopia of a musical philo-fiction for generic humanity.

PREFACE TO
THE FRENCH EDITION (1992)

THIS book is a contribution to a few local problems that philosophy is currently examining, like "singularities" and "fractals," but above all to fundamental problems of philosophy *itself*. It does not comment on texts; it forges new tools for thinking at last, in global terms, and for transforming the theoretical status of the philosophical genre. It does not question yet again its origin and its historiality, but asks how it can afford philosophy a future other than that of memory and nostalgia, commentary and deconstruction. Under what conditions can philosophy still be useful to us without being, as it has always been, conservative and authoritarian? True, the philosopher is a hero, but he is a fatigued hero whose life is a life of survival and whose vigor is the vigor of sudden fits and starts. How can we make this discipline enter into the concert of the sciences? How can we finally satisfy this requirement of reality and rigor that philosophy has only ever been able to half-fulfill by its own means, but without reducing and weakening it in a positivist way? A scientific reform of the philosophical understanding—such is the program that four fundamental concepts punctuate: identities-of-the-last-instance, nonepistemological conception of science, generalized fractality and chaos, and artificial philosophy.

The central problem is the one through which contemporary philosophies have critiqued and "liquidated" the Hegelianism, the xviii • Preface to the French Edition

Marxism, and the structuralism that preceded them: the problem of *singularities* and *differences, partial objects* and *critical points, catastrophes* and *effects, disseminations* and *language games* . . . All of these objects were directed against "logos," "presence" or "representation," "metaphysics" and so on, and have become philosophy's commonplace. But from our point of view, they represent a half-solution, an unfinished attempt at the critique of metaphysics. Why? Because they always associate with these singularities of various types an *identity*, but *as at least equal to them or reversible with them*. Identity thus falls back on the singularities, appropriates them, capitalizes or traditionalizes them, subordinates them to an indeterminate generality, and so forth. This solution entails their erasure or drowning.

At their origin, nevertheless, those objects were scientific and belonged to thermodynamics, differential calculus, the theory of sets and of critical points, the theory of catastrophes and of fractals . . . But their appropriation by philosophy (ontology, Kantianism, Nietzscheanism, structuralism) and their placement in the service of Being, Desire, and Language contributed to limiting their scope. Their philosophical generalization partially effaced them.

To restore to these singularities their positive and critical vigor, their theoretical dignity, we propose to reinscribe them in the content of science rather than of philosophy; but the content of a science that is itself rethought and described in its essence in a new way. This new description is no longer epistemological, i.e., philosophical—it is in fact futile to try to free multiplicities and singularities from their servitude without freeing the science that produced them—but properly scientific. In the wake of earlier works, we seek to demonstrate that science also "thinks"; that it is a specific and original way of relating to the real, distinct from the philosophical way; that it can thus describe itself. We call this conception *epistemic* and no longer *epistemo*-logical. This is why the book opens with a systematic exposition of science's "nonepistemology". The outcome of this description is that science and identity entertain the most intimate relations; but this identity no longer has its traditional essence (transcendence) nor its philosophical functions (totalization and closure). We call it real Identity or Identity *of-the-last-instance*. This formula is to be taken, in its strict theoretical or nonphilosophical sense, to mean that Identity is not alienated in that of which it is the Identity, in its effect, and correlatively that it autonomizes this effect without folding back on it or reappropriating it.

Thus understood, this Identity emancipates singularities that are at last radical, fractals that are no longer subjected to it. It gives them their *reality* and prevents them from dissolving in philosophical possibility. It authorizes the constitution of an autonomous order of singularities in the form of chaos. We will describe this nonepistemological generalization as simultaneously scientific and not regional (geometrical, set-theoretical, catastrophist, thermodynamic, etc.); transcendental and not philosophical (Aristotelian, Kantian, Nietzschean, Lacanian, etc.). It alone "conserves" the singularities in a nonconservative way and imparts to the old, rather worn-out philosophical "multiplicities" a new power, which will be the power of generalized "fractality" and "chaos."

This contribution to the theory of a few contemporary problems presents a precise difficulty for the reader. It is a matter of a new type of intersection between science and philosophy: a nonepistemological inter-section. For us, it is no longer a matter of philosophically reflecting once again on science; but of conjugating recent scientific objects (fractal objects and chaos) with philosophical objects, which are rethought in a new way: Identities—as what is no longer their "foundation," which effaces them, but their "cause of-the-last-instance" that safeguards them. This is the source of our text's peculiar difficulty, which we cannot conceal. Neverthe-less the difficulty resides not so much in the expression as in the nature of the examined problems and objects: it is "objective" and has to do with the content. We ask the reader to penetrate into a manner of thinking that has both scientific traits (yet without mathematics, without "equations," always in natural language) and philosophical traits (but also critical of philos-ophy). To mitigate this difficulty, the book is progressive; it proceeds by increasing complication, by a continuous introduction of new objects. But above all it has a "fractal" nature: each chapter reexposes in a distinct mode, at different "scales," and under variations of objects the same structure of nonphilosophical inequality or irregularity, of fractality-in-philosophy itself and no longer in "logos," in "presence" or "representation" . . . Thus by osmosis, by habituation to invariants, the reader gradually penetrates into this manner of thinking.

The problem of Identities and of singularities serves as our guiding thread, and the theoretical status of science and of philosophy is examined and formulated along the way, but three concepts or three theoretical dis-coveries are the direct objects of our research.

1/ The first concept is *Identities-of-the-last-instance*. It renders possible *a theory of science that is no longer epistemological in nature or philosophical in origin.* This concept cannot in fact be discovered—at least as it has been expounded—in traditional philosophy, including deconstructions (Wittgenstein, Heidegger, Derrida). It serves here to describe the chaos-essence of the real and that of science.

2/ The second concept is *Generalized Fractality, intended to replace the concepts of "singularities" and "multiplicities," "differences" and "disseminations."* Taking as our guide or theoretical signpost B. Mandelbrot's works (which have been universally accepted by the scientific community and widely used in various areas of research), we generalize them in the aforementioned mode, which remains internal to science, but to a science endowed with an authentic power of relating to the real (Identity) and of thinking this relation. *Generalized "non-Mandelbrotian" fractals are then no longer objects of "nature" but of knowledge or of theory, and above all they form a novel theoretical tool, adapted at last to the disciplines of language* (philosophy, poetry, literature) and no longer only to geometrical and perceived forms (physical phenomena of turbulence, cartography, painting, photography, etc.). This concept is accompanied by others ("fractal a priori," "fractal intentionality," "generalized chaos," etc.) which cash out its theoretical force in a nonpositivist mode.

3/ The third concept is *Artificial Philosophy*. It responds to a project and to a solution. The project: can an Artificial Philosophy be established in the mode of "Artificial Intelligence" and certain "cognitivist" practices, or practices that can be envisioned within the framework of a "philosophy of the mind"? Can a synthesis of philosophical statements be imagined that retains a philosophical or nonpositivist *type* of value, but that is realized by means of science? The solution: if it is impossible to computerize philosophy itself without destroying it as philosophy or destroying it in its "transcendental" dimension, without reducing it to local and abstract, mathematically dominatable problems, it is, on the other hand, possible to take another path that uses theoretical means of a different nature: no longer mathematical but linguistic and in some sense "qualitative." *The deciphering of philosophical texts or problems by the rules of a generalized "non-Mandelbrotian" fractality, rather than a geometric fractality, allows the synthesis of artificial statements,* which have one last philosophical color but no longer respond to philosophical codes of acceptability and admissibility. Such synthetic statements are "artificial" in a more powerful sense of the term than simple "artificial" Intelligence. For they are produced on the basis of sciences' transcendental essence rather than on the basis of this or that particular science or information technology. In general we call "non-philosophy" the type of activity that uses philosophy under scientific and no longer philosophical conditions. And if there is a philosophy as "rigorous science," it will instead take the form of a *science of philosophy* or a *non-philosophy*. The Idea of an Artificial Philosophy is only one of its particular but nonexclusive modes of realization. The ultimate goal—if it is still a goal—is to "apply" a theory of generalized fractality and chaos, which does not itself have a philosophical origin, to philosophy, to its most invariant decisional and positional structures. From this point of view, our entire book is a search for conceptual formalisms and rules of theoretical treatment, of algorithms, which allow the use of philosophical statements in view of the production of synthetic statements. These statements make possible a better analysis and a more radical critique of philosophy; they represent, more positively, a crossed threshold, a mutation in the traditional exercise of thought.

Philosophy's future—if we wish, of course, to imagine and realize such a thing—surely does not lie in its infinite commentaries or its interminable deconstructions, and even less in its naive, more or less positivist practice. It lies in its intimate, nonhierarchical cooperation (once all its legislative ambitions have been deposed) with sciences, according to relations of use that exclude every will to domination. Something different, therefore, from a "new alliance." This research would like to have contributed, however slightly or in a very elementary way, to the global reevaluation of the relations between science and philosophy, to the destruction of their *unitary* (epistemological or philosophical) *theory,* and to the establishment of a *unified* (scientific) *theory* of thought.

THEORY OF
IDENTITIES

INTRODUCTION

SCIENCE, IDENTITY, FRACTALITY

THE ERA OF MULTIPLICITIES
AND THE FORGETTING OF IDENTITIES

CONTEMPORARY philosophy has put an end to Hegelianism, Marxism, and structuralism by drawing our attention to new objects of a special type. Although very heterogeneous, these objects are globally foreign to "metaphysics," "presence," and "representation": partial, fragmented, irregular, or fuzzy objects; badly assembled, mismatched, or dehiscent apparatuses; multiplicities and disseminations; games; differences and differends, in-betweens, and so forth. Since Friedrich Nietzsche, philosophy has been mobilized in a struggle against "logos," understood as *unity* ("system," "representation," "hierarchy," "closure," etc.). This struggle has been the source of an inflation, a bidding up [*surenchère*] that has lead to the intensification of "multiplicities," "inconsistent multiple," "language games," "catastrophes," "turns," "effects," and "singularities." These objects engender affiliated theories. Competing with one another, they compete even more, all together, with the equivalent scientific forms (critical points, bifurcations, catastrophes, fractals, and so on) to which they correspond and of which they are perhaps, simultaneously, the philosophical *aftereffect* and rival.

Nevertheless, those philosophies—and perhaps all philosophies without exception—quickly came up against typical difficulties. They sought in the invention of the new objects a "postmetaphysical" practice. But in several ways they merely extended through them a certain metaphysical impotence of philosophy and repeated on them their most traditional errancies. We shall return to these difficulties, which testify to the unreality of the aforementioned objects, to their merely possible character. But here they are, in short:

1. Those objects are not real or consistent objects as defined by sciences, but created or imagined artifacts, "decided" at will; *mixed*, half-given half-decided entities. Their degree of reality outside philosophical practice is and has remained entirely problematic.

2. They are of multiple types or genres at the whim of competing philosophical decisions. Their sole reality is that of philosophy itself, and philosophy behaves toward them as a tradition or a "reserve" of thought to the teleology of which they are ordained. It puts them to work for the benefit of its own goals or objectives, extracting from them a surplus value of sense, of truth, and of value.

3. They have no stable identity. More precisely, they are directed against every identity. Here *identity is mistakenly conflated with a homogeneous and transcendent unity that would circumscribe them*. They are caught in becomings, lines, textual or other interminable processes, in which they are exhausted. Their reality as points, terms, or individuals is drowned in tendencies, continuums, and infinite teleologies.

Our general thesis on this point is the following: "singularities," "multiplicities," "differences," etc., 1/ exploit transcendence exclusively, profit from Being and exteriority, are the avatars of ontology and its power of *autodissolution*, which is at last manifested as such; 2/ project in their turn a transcendent or ontological (thus negative) image of the One or Identity in the form of Unity, with which they conflate it; 3/ requisition Identity, i.e., *the One as One*, without having elucidated its essence, its nature as a Given, which they negate by confusing the One with the Unity-for . . . ; Identity with the Identification-of . . . , etc.; 4/ are mixtures of Identity and of Division or Scission; of false Identities and of true generalities, *partially indeterminate and transcendent*, which deny the real's authentic "singularity," conflating it with conceptual singularities, blending it with philosophical representation. Their element is dissolution, dissemination, deconstruction . . . the weakened forms of the negative. They are the excess of transcendence over itself, a product of its autocritique rather than the recognition—positive and for itself—of the *identity* (of) the real. So much so that, since every singularity is always too weak and presupposes another singularity, this thought culminates in a hermeneutics of philosophical texts and in a homage to Tradition. They are what remains of the traditional, unitary metaphysics when it undertakes to critique itself and continues to impose its oldest prejudices.

To be sure, contemporary philosophy has partially understood that, in order to ensure their *reality*, it had to cease recapturing singularities in a foundation, inscribing them in a representative generality, and had to ascribe them to the Other-of . . . or the Other-than . . . But it merely completed half the path, the less interesting half, the one it could complete by remaining "philosophy." It transferred its old habits of self-sufficiency to the Other; it even accentuated them, by assuming this time Exteriority itself, Transcendence, the Other, as an Absolute in a state of "autoposition" or "auto-affirmation," certainly attenuated but always arbitrary as well. It associated singularities yet again with a kind of universal and all-consuming Reserve (*Episteme, Full body, Plane of immanence, Place-for-the-Other, Clearing, Tradition,* and *Destiny,* etc.), a reserve that has exploited them with the good faith peculiar to cynicism. Philosophy counts on the infinite unlimitedness of time to save the singularities. It refuses to acknowledge that this infinity of philosophical time is a "bear's service" [*le pavé de l'ours*], that it is thought-as-Reserve determined to exploit these singularities. It cannot help but "possibilize" them, turn them into semieffective, semipossible, semivirtual, semiactual mixtures, *divide* them, and destroy their identity. This identity is lost with all hands *because it is a priori or by right "mixed" with that of which it is the identity.* So much so that these singularities' transcendent mode of existence (worldly, social and historical, textual and linguistic, desiring and political existence) drags their identity in an endless turnstile and their reality in unlimited games of the most decisory [*décisoire*] and arbitrary possibility. It is not surprising that these supposedly "nonnegotiable," "incompossible" . . . objects have been used to negotiate the maintenance of the oldest myths of the philosophies of history. They have served as an ultimate caution to philosophical sufficiency, a new mask for its claims over the real.

Clearly the revival of the problem of "singularities" and of "multiplicities" requires us to change our foundation—but what does this formula mean?—and to turn toward the original problem of Identity so as to give it an *other-than-philosophical* form, capable of affording singularities a reality that philosophy has contrived to strip from them.

PHILOSOPHICAL RESISTANCE TO IDENTITIES

Let Difference and Identity be "transcendentals" in the contemporary sense, absolutely general "metaphilosophical" categories that allow us to condense a philosophy or philosophy itself into a formula, for example: "Identity of Difference and of Identity." Their circle, co-belonging, or reciprocal determination distributes the two terms unequally: Identity appears dominant since it intervenes twice: once as a simple term, another as a unity of the system of the two; as party and as judge, as productive and as contemplative. Difference only intervenes a single time, as an operative term of the whole.

But this is just an appearance that can mislead us as to philosophy's sense and goals. In reality, the formula means that philosophy's primary and dominant element is Difference. Let's say the "Dyad": Difference is not the Dyad; it is an already specified mode of the Dyad, which is a more indeterminate matrix. Difference is Dyad and One, at any rate, a particular philosophical system. But it can also represent the dyad side of every philosophy, and it is possible at a certain level of generality to assimilate the two. Philosophical thought begins with difference—with the multiple, experience, decision, etc.—*and moves within this element when it thinks Identity*. Identity is divided into several species or modes, so much so that its double intervention is the mark of its weakness; its final overvaluation is the effect of its initial undervaluation. Philosophy *exploits* Identity as a keystone of systems, but does not think according to Identity. Having divided it, philosophy resorts to palliatives or substitutes: analogy, univocity, and so forth. Identity thus forms two circles: the small circle that opposes it to Difference, to the Dyad, but that nevertheless enters with the Dyad into a superior Dyad; and the great circle through which it surpasses itself and reenvelops Difference or the Dyad it constitutes with Difference.

Difference is therefore the essence or motor of philosophy, Identity its existence and objective appearance. This hierarchy is complex. It separates the functions, distributes the roles, and decides the share of each. In effect, philosophy is a reunification only because it is and remains a *transcendental partition*. It is *in a dominant way* an activity of decision and determinant division from which thought never exits. Even when it aims for Identity, philosophy still thinks it in the last resort in the mode of its division. Even when it assumes Identity in an elementary way, as *the principle of identity*, it is never A, but A=A. There is no philosophy that does not maintain a *double*, even multiple discourse on Identity: that of the numerical unit-of-account, the atomic unit, an ingredient of arithmetic or of the sensible multiple, but also that of transcendental Unity as an ingredient of Being and convertible with it (or with the Other in contemporary deconstructions).

This savage will to the "forgetting"—the denial—of Identity takes several forms on the basis of its preliminary division. The most noteworthy is the form of its convertibility at times with the numerical *one* of arithmetic, at others with dominant Totality or systematic closure—in all cases, ultimately, with . . . Unification. In each case, its convertibility replaces its real essence and devotes it to "technological" works that buttress the ontological decision. Unable to rid themselves completely of the One (which is necessary for arithmetic, but more profoundly for the relation of the multiple to the One itself), some philosophers fantasize its murder. They fantasize the liquidation of the "Greek god of the One" and believe they "make the decision" by deciding to invert the old hierarchy: $\frac{One}{Multiple}$ into $\frac{Multiple}{One}$. In reality, they merely "decide," i.e., fantasize a murder that has already occurred a long time ago, a murder tethered to philosophy's very existence rather than to their own decision and which, if it testifies to a redoubtable, murderous will on the part of philosophy, is a murder for laughs with regard to Identity itself. These philosophers claim to have finished devaluing the One as a simple unit-of-account and to have overturned its old metaphysical empire in the name of an uncountable multiple. Such a division allows them, on one hand, to reduce it to the secondary functions of unifying the Dyad or consistently closing Difference; on the other, to accuse it of all possible evils (closing, systematic closure, domination, totality, etc.). In these accusations, philosophers are manipulated by philosophy, which is nothing more than the unity of this double gesture of devaluation and overvaluation. Those who claim to put the One to death continue to secretly overvalue it, for they think inside a system that desires these two gestures at once. No philosopher can do without the One and its unifying functions, which are *always surreptitiously presupposed*. No philosopher can avoid limiting its efficacy. For philosophy, Identity is a necessary evil. We cannot believe that it is or has ever been philosophy's main object. It is Being that is its primary object, and this is why philosophy overvalues the One: because it seeks its ruin. The idea of overvaluing and devaluing the One, i.e., the real, is absurd. It is, nevertheless, the foundation of philosophy.

Difference can only be understood as a form of amphibology, of forced unity between heterogeneous terms, of apparent identity between terms whose own nature or real identity renders them foreign to each other. Conversely, the amphibology claims only to divide the real identity of these terms. But it is the whole of philosophy—pervaded by Difference of which it is the development—that is the activity that puts Difference where Identity is and Identity where the true "difference of nature" lies, the heterogeneity of Identities. Difference supplies the purest philosophical, i.e., amphibological satisfactions: *two becomes one* at the same time that *one becomes two*. Philosophy is not only Difference's *priority*; it is its *primacy* or its *domination*. When Difference is primary and when one begins by deciding, dividing, partitioning, or simply distancing, this gesture engenders universal hierarchy and domination as the essence of thought. Even when it apparently cedes its place to Identity, this place continues to be that of domination. Contemporary philosophy, at least the philosophy of "deconstructions" (Ludwig Wittgenstein, Martin Heidegger, Jacques Derrida), has not eliminated this primacy of difference; instead, it has intensified it, injecting a supplement of alterity into Difference so as to prevent its perfect or remainderless interiorization into Identity. Plugged into an Other, which is doubtless not the form of an Identity's scission but the power of the scission, Difference was "opened" or quartered. It has become an irreducible distance, an irreducible tension. For this reason, deconstructions have not destroyed philosophy; they merely reinforced its dominant internal tendency: the dislocation of Identity. They took this destruction of the One to its limit, which obviously proved to be a limit of impossibility. *There is (always) some One*—and not only as a unit-of-account—*even and especially when the One is presumed to be expelled from the "real."*

On the basis of this description, the following hypothesis can be proposed: philosophy as a whole is "opposed" to Identity, since it divides it and fails to think it *as Identity*. The entire philosophical activity of differentiation and identification (dialectic, synthesis, system, arithmetic, etc.) relies on this global rejection. That of which philosophy speaks and which it assembles with Difference (with Being, the Other, the Multiple, etc.) is merely an *image* or an *appearance* of Identity (under conditions foreign to the latter), but is not Identity *as such*. Philosophy can in no way think Identity and only gains access to it by dividing it, reducing it to secondary, even "vulgar" and "empirical" functions (arithmetic) and simultaneously raising it to the state of a transcendent absolute beyond Being and in positions of domination it immediately proposes to "critique," "overturn," or "deconstruct." This explains a fundamental trait of philosophical activity: Identity qua Identity—and the science of Identity—is a formula that we can discover in ontology; but it is never fulfilled, no more than any formula of this kind. Philosophy is not a science of the One but primarily of Being, a science that tries to find its bearings, fails and elevates its failure to the state of success, an aporetic science that establishes itself as a "science" of the aporia, a sought-after science that becomes a "science" of the search.

Once the extent of this repression (and, in the first place, the depth of the problem of Identity) has been grasped, it would not be an exaggeration to say that philosophy only lives off its resistance to Identity, its denial of the One for the benefit of Being, that it starts to "speculate" about a transcendent, ontological One or, weary, reduces it to the unit-of-account. Philosophy's global failure "before" the real stems from the fact that it thinks in the mode of the *decision* and, more generally, of *transcendence*, already beyond and against Identity, and that it only seeks the reality of singularities in singularities themselves, *thus* in the transcendence that effaces them, in the exteriority that dissolves their essence. Philosophy has only ever possessed a superficial, transcendent concept of Identity. In the One, it discovered a kind of rarefied double of Being. It is the system of a superficial experience and of a naive critique of Identity.

The balance sheet of the philosophies of "difference," of "multiplicities," of the "multiple," and finally of "deconstructions," is what we could call a *semidiscovery*, an incomplete discovery of singularities, or more than incomplete: falsified and symptomatic from the start. "Continuous multiplicities" fated to the erasure of their identity, a mixed semisingular semiuniversal real, and possible inasmuch as it is real: such is in effect the meaning of this ubiquitous appeal to difference, to the differend, to the inconsistent multiple, to language games and to catastrophes. The *science of singularities*, in the same way as philosophy and for the same reasons, remains a "sought-after science" and will eternally remain so as long as the very problem of science (to begin with and in its relation to the One rather than to Being), not to mention that of multiplicities, is posed under the philosophical horizon.

CHANGING THE TERRAIN: SCIENCE AND SINGULARITIES

The events that haunt the World's surface may be objectively narcissistic. Perhaps they themselves appreciate this labor of division and envelopment to which they are subjected—care and attention of philosophers. Nonetheless, it is not clear that those who engender events or manifest them (artists, engineers, scientists, or any man insofar as he creates, believes, imagines, legislates, writes, crosses space, saves, or kills) equally appreciate this labor of assimilation and recuperation. Events have no shortage of all kinds of philosophies and interpretations; in fact, they have too many of them. There is an objective excess of philosophy or of mediatization, which ends up concealing singularities' virulence and fulguration.

If philosophy is the mistress of this enterprise of *drowning* determination under the triple veil of sense, truth, and value (the three dimensions of space it folds onto the real), we maintain, in contrast, that artists—but this is true first and foremost of scientists—have already won their majority, that they have escaped the state of minority in which philosophy held them captive. We no longer believe that philosophy represents the "majority" for man. Rather, precisely because it is merely a majority of the understanding or of reason, it constitutes the reign of the imposed minority. It often holds artists in a state of dependency, even though scientists have already broken with this servitude. To be sure, when artists cease in their turn to be the minors of philosophy, it is then that they enter into the community of true "minoritarians." The authentic minorities, those who no longer let themselves be identified and determined by philosophical Authorities: this is what the very "experimental" research undertaken here will try to discern.

The old problem of "determination" and the most contemporary problem of singularities constitute a possible entry point into this new way of thinking. A historically dated entry point, no doubt, and required by the urgency of situation. But it is the global work of scientists, artists, and technologists that we should take as our object and not only that of avant-garde philosophers, at the risk of losing the generality of the paradigm we call "nonphilosophical" and covertly returning once more to a shameful philosophical decision. That the events, objects, and procedures of the hour appeal to us more than others, that they become a symptom for us, is perfectly reasonable, provided we distill each time the scope and radicality of the new paradigm, inasmuch as it can be measured against philosophy's paradigm and, more than measured against it, become a measure for it.

In effect, it may be more interesting to understand the work of artists as a set of procedures designed to demediatize the event. Demediatization of the real rather than deconstruction of interpretation; production of a naked singularity from clothed singularities; subsumption of envelopes associated with the event into the materiality of the event. Still, the expression *demediatization* betrays what is at stake and must be "corrected" and made possible. Instead of suspending interpretation in order to reascend toward a supposedly pure real, toward a disrobed singularity—the old and circular philosophical schema of suspension or reduction—we would start from the already-Determined in order to move toward philosophy's folded singularities. We would treat these singularities as a simple contingent material on the basis of rules or procedures that express the real's unreciprocated precession on philosophy, singularity on its interpretation, the Determined on Determination. We would thus circumvent not only the philosophical gesture of retrocession or ascent toward the real, which is illusorily presupposed outside philosophy or on its margin, but also the trap of a primary empiricism or of a materialism of materiality.

What conditions produce a radical singularity, not enveloped or leveled by philosophy's transcendent spaces? To this question, contemporaries responded with the prodigious recourse to the *Other* in increasingly heterogeneous and undecidable forms. This solution is conservative; it does nothing more than register what is in question. By displacing the philosophical operation toward the nonspace (of) the Other, it simply confirms the supposed necessity of the mixed or philosophical state of singularity instead of positing its contingency and its unreality. Hence the invasion of thought by these effects-(of)-the-Other: the symptom, trauma, heterogeneity, the fold, fragments or fractures, catastrophe . . . all that constitutes the multifarious content of the "postmodern" (understood more or less rigorously). Isn't it instead the question itself that is conservative and that already programs the division and doubling of singularity? How can we claim to produce determination *ex nihilo* or even from what is not already the Determined and what contains some indetermination in itself? This impossible task is that of philosophy. The real work of scientists and artists is elaborated according to a whole other axis. What is at stake—if we treat the already-Determined as a guiding thread of the description, but also of the real operation that scientists and artists carry out, which is neither an interpretation nor a deconstruction—is: 1. *extracting* from enveloped singularities their *real* or phenomenal content, i.e., a singularity stripped of every decisional and positional environment; 2. *overdetermining* it, but from now on without dividing it—by means of mixed singularities that are interpreted, enveloped with decisions and positions. Two simultaneous operations, whose sum is the transformation of these clothed singularities in accordance with the Determined, not the production, which is at any rate impossible, of the Determined itself. It would be naive to claim to negate the systems of interpretation that accompany the real of the World, of Art, of History, and of Politics in the name of an allegedly purer real. Our project does more than reverse this path: instead of allowing philosophies to exploit the *reality* of singularities, we make use of them as resources in the service of this reality. What we reject is the autoposition of these interpretations, which claim to occupy the entire space of thought.

It is at this nonphilosophical level that the problem of a more radical apprehension of singularities must be posed, not at a socio-semio-etc.-logical level, which is merely a disguised philosophical position. Philosophy is an enterprise of idealization and of composition—a technology—of singularities, but this expression contains a paradox. Can't we imagine another paradigm of thought *that would no longer associate the ideal or universal side of singularities and their empirical side, but would elaborate them according to wholly new nontechnological relations, outside every causality of division/doubling and with a greater respect for their specificity?* The philosophical conception does not only lack the empiricist sense of what is given as such, contingently or "in excess," the sense of the World's unlimited horizontality or its "flattening" (the empirical is necessary to philosophy, yet it is also intolerable to it; it "inter-venes" ["*inter-vient*"/inter-comes] in it). It also lacks the sense of ideality and of functionality. It is neither quite Humean nor quite Platonic, but carries out a *blending* above all. What is singularity in the philosophical sense? The blending of the empirical manifold and the ideal unity, their co-belonging, with the movement that animates this blending. Thus: their interimpediment or interinhibition. Can we imagine another way of thinking for which singularity, which certainly always constitutes a complex edifice, would no longer have the internal structure of the interinhibitive mixture or circularity? A double attention, perhaps a double donation, is required to do justice to singularities: the sense of the absolutely contingent and nonstandard given, treated as such and not transformed, the sense of the theoretical ideality that is independent of every empirical content. If these two dimensions are no longer *associated* in a mixture, what will be the law of their new relation? Perhaps it is necessary to defy the spirit of blending to the point of positing a duality—radical, unheard-of, precisely without blending—of the real and ideality and, on this basis, to discern how both can still, if not "compose," at least produce a singularity of a new type. To singularity's mixed and auto-effacing state, we will oppose its double donation $\frac{D}{\frac{d}{d|u}}$ where D is the radical Determined (Identity), d the mixed determination or singularity, and u the universal or the space associated with the singularity. A nonphilosophical, perhaps scientific redistribution—founded directly in the real itself—of the Determined, of Determination, and of the Determinable: this is where we hope to locate the new resources for a more faithful and rigorous description of the materiality of the World, of History, of Science, and of Art. These various fields contain an ingredient-of-reality that must be described and that can only be described if the phenomena they offer stop legislating on themselves and self-interpreting and are treated as a pure inert determinable ready to receive the procedures of determination that flow from the real or from the Determined itself. To elaborate this new concept of singularity, we have to return to the old problem that serves as its framework: the *Principle of Sufficient Determination*.

We will therefore attempt to elucidate—on the basis of the symptom-problem of determination, of singularity, and of fractality—the absolutely primitive and minimal structures of every thought. These structures will be reduced and simpler than those of the philosophical decision. Far from representing a new "return" to things themselves and to a metaphysical "simplicity," or a simplicity of the retrocession-beyond-metaphysics, this attempt realizes an antedecisional-and-positional simplicity and minimality. It is science that possesses these incompressible structures of every thought, and not philosophy, which merely lays claim to the "inescapability" of its own existence. It is from this essentially indestructible or irreducible (because real) given that we set out to describe sciences, arts, and even philosophical "complexity." Philosophy's infinite and wily developments, its overcomplexity, not to mention the "theories of complexity" that are its ultimate extension, create the illusion that philosophy is capable of representing and describing a "real" deemed to be increasingly "complex." Yet, since philosophy itself starts with complexifying the real by investing and projecting itself specularily in it, we find here an eminently vicious illusion and a vicious solution that conflate the real with "reality." Perhaps complexity is only accessible to a thought that will not have retroceded once more to a "simple" metaphysics or to a metaphysics of "withdrawal," a thought that will have assumed the means of *starting* from philosophically unheard-of simplicity and minimality: those of the *scientific posture* rather than those of the *philosophical position*. So as to summarily extend this type of distinction, we could oppose the revolutions of the complex or the philosophical and the mutations of the simple or the minimal; the philosophical's internally closed openings and the scientific's radical openings; the increasing philosophical complexifications and the scientific transformations-mutations of the complex (representations or knowledges) under the law of the scientific posture "in" the real. We have to develop these aspects of the "scientific" or "nonphilosophical paradigm" and elucidate it in terms of the crucial problems of determination and singularity.

Singularities must stop being a "question" in order to become a theoretical "problem" posed on the terrain of science. But science must also become a problem posed on the terrain of thought. It is a matter of *discovering a rigorous form of thought, with the traits of science rather than of philosophy, capable of affording the philosophical half-discovery of multiplicities its true theoretical sense and the plenitude of its reality*. To save singularities from their philosophical effacement through their theoretical implementation: such is the "watchword" that we baptize with the term *nonphilosophical*. But the paradox is that, for this operation, we must agree to give the *appearance* (in the eyes of the philosopher alone) of a kind of "flashback" toward the problem of Identity, which is particularly repudiated by contemporary philosophies as well as poorly understood or repressed by them.

GLOBAL REEVALUATION OF THE RELATIONS BETWEEN SCIENCE AND PHILOSOPHY: STARTING FROM "SCIENCES THEMSELVES"

Measured against the obsessions of our time, against its objects and criteria, what we call "nonphilosophy" is a theory of identities rather than a philosophy of singularities. All the same, we do not believe that it marks the return to a metaphysical state of thought. Rather, what is stake is its displacement, its *mise en place* [establishment], on the terrain of science. Not the terrain of any (philosophical) image of science whatsoever: the fundamental task of the solution on which everything will depend consists in knotting (and more than knotting: identifying) Identity qua Identity, finally restored to itself, and the essence of science. We will have to describe this radical imbrication of Identity and of thought-science.

Sciences are everywhere triumphant, to the point where they are undisputed in their sphere and are rejected solely out of "ideological" or philosophical distaste on account of "culture." They triumph as the sole incumbent body of theoretical knowledge, in possession of the conditions of historical change. An observation: philosophy itself recognizes that it has lost the great battle of theory. A paradox: science in no way has the thought it merits; it is left up to its philosophical consumption.

On one hand, philosophy has lost in fact—and recognizes this loss through its avoidance and cultural consumption of science—the great combat it always waged against science for the control of theoretical knowledge. *Always*: even when science did not exist, it already occupied the future battlefield. But philosophy renounced theory and explicitly took refuge in unending side tasks or secondary ends: everyday practice and the maxims of prudence; the "resistance" that determines the choices amid sciences and powers; the quest for happiness and the new "honest," i.e. mediatic, culture. New consensus where everyone must be recognized: companies, educated men, educational and academic institutions, scientists who are asked to expound their "philosophy." Philosophy returns to a life of misery by visiting the scene where it is summoned as a "noble" force of conservation and "resistance"; reduced to the functions of self-defense and security; in charge of expressing the distaste of "intellectuals" against the real labor of sciences, against theoretical dignity. It is true that it never thought *in order to think*, but in order to live or survive: as an aid to life and the decision. Philosophy is in its essence a *practice* of prudence with *aspects* of thought; it is not a practice of thought, but a desire for thought that gradually recognizes, in this respect, that it is merely an aborted dream within the dream. If it is also something else, it obtains this extension from science.

On the other hand, despite their triumph—or because of it?—the sciences are awaiting a *thought adequate to theoretical knowledge*. Instead of this *thought of science*, we have the rearguard battles that end in the "returns-to . . ." (to Immanuel Kant, to Leibniz, to Aristotle, etc.) and the epistemological self-defense groups. We have the obsessional return of the old solutions—those that were forged in response to prior states of theory and knowledge. We have the epistemological consumption of knowledges, isolated and abstracted from their process of production. We have the "philosophies of science" or those founded on an old scientific theory and that, by definition, lag doubly behind: behind the "current state" of knowledges in the field and because philosophy is late de jure, displaced in relation to science, in a "reactive" position. We have ideologies that scientists project through their work. More generally, we have the globally specular image that philosophy cannot help but ascribe to Science [*la* science], the unitary confusion it cannot but introduce between the two of them, the exchange it requires between "knowledge-without-thought" and "thought-without-knowledge." The founding confusion of our time, immemorial like philosophy itself, is not that of Being and being; it is the confusion of science and philosophy in the name of philosophy against science.

This will be one of the tasks of the future historians of spontaneous philosophical representations (let's try to imagine them): to describe the forms of this extremely protracted repression, the strange strategies through which philosophy will have passed itself off for so long as the authentic thought of science when it merely reflected or expressed itself, without restraint, in the distinct, ever transcendent images it created of science. It will have incessantly imposed limitations on science, endowing it with foreign objects and external goals, derealizing it in order to set itself up as science's real foundation. In general, philosophy will not have allowed science *to think on its own*; it will have put itself in science's place in order to speak "in its stead" and to expel it from itself, to alienate it.

The first task is thus to set out from what we will call "sciences themselves." This will be our "watchword" and no doubt a whole other thing than an imperative or an injunction. Our *axiom*, from which we deduce science's essence, scope and autonomy vis-à-vis philosophy, is *science thinks*; it thinks without philosophers and despite all the apparent counterexamples. Instead of understanding by "science" a particular and complete knowledge that philosophy abstracts from its element, appropriates and "autoposits" as a fact (*faktum*) or an "object," we will understand "science itself," its *identity*, which can think itself in its philosophically unprecedented way. Everything will then mutate in the specular economy of philosophy's relations to itself: the "philosophy of sciences," epistemology, and scientific ideologies will reveal their nature as activities of cultural self-defense.

"Nonphilosophy" is the attempt to elaborate a *thought* that responds to the most general scientific criteria (internal or "transcendental" determination, rigorous coherence, exclusion of circularity in favor of deducibility and experimentation, etc.). It is thus the attempt to elaborate a thought adequate to the theory and immanent practice of sciences. It is a theory of the *essence* of theory. Yet the theory at issue not only excludes positivism but also idealism and theoreticism, the philosophical *normalizations* of science. Theoreticism in particular would arise if, in isolating *a* theory, we assumed that it autoposits and autolegitimates itself and "gives" the real, as philosophy assumes. Here theory is determined in-the-last-instance in or by the immanence of real-Identity. This *cause-of-the-last-instance* of theory shatters every theoreticism and liberates all the more the theoretical dimension from its (philosophical) confusion with the material or the object.

What troubles philosophical common sense is that nonphilosophy is simultaneously, in its practice and not only through its object (philosophy), *a science and a thought*; that it has scientific rather than philosophical operations, and yet they are a priori and transcendental; that the two traditional, opposed images of so-called positive science and philosophy are no longer valid, that these old partitions are abolished for the benefit of the *identity* of *science* and of *thought*. An identity that rejects both the positivist image of science and the reflexive (autopositional, redoubled, or folded . . .) image of thought; that rejects them not merely for its own good, but for the good of science as well. On the whole, this schema is closer to the spirit of "sciences themselves" than to philosophy, i.e., to the unitary or hierarchical opposition between sciences and philosophy. Our constant image of the sciences has a philosophical origin and must be eradicated by sciences "themselves."

We no longer believe that what—perhaps for all eternity—is dubbed "philosophy" could still claim to supply thought's ultimate structures. Philosophical Authorities decide and legislate; they manifest a vigor and an insistence that forbids any "decision" on their inexistence. On the other hand, beyond their effective existence, they do not cease struggling against, not a mere suspicion, but an absolute rejection that is opposed to them everywhere and which, in order to defend themselves, they mistakenly reduce to a positivist bad will or an empiricist obstinacy, to a scientific barbarism or an anticulture that besieges them from all sides. The real situation is simpler, or else more complex. Perhaps philosophy is the one that has not ceased repressing and resisting another way of thinking that is now emerging, *that has already emerged and claimed the plenitude of its rights without the slightest awareness on philosophy's part*. What philosophy takes for an external crisis, an assault on "free thought" and "reflection," what it seeks to turn into the deconstruction of the most metaphysical forms, are perhaps the cracks in its edifice, the fissures in its own resistance. A certain becoming of thought is delineated, a history is discerned, but it is less the emergence of another thought inside and outside philosophy's limits than the autodissolution, the autocritique by which philosophy reacts to its basic impotence in the face of what has always been there "before" it and what emerges a little more distinctly through the invalidation of its claims. As old as it is, philosophy arrived too late in relation to another way of thinking, which can be primarily identified as that of science, then as the one that acts, for instance, in the work of artists. That there is an authentic and consistent thought in science, that it does not depend on external and, in particular, positivist and empiricist philosophical decisions, is precisely what we call *the scientific or "nonphilosophical" paradigm.*

The new problem is then no longer the fulfillment of philosophy, its minor positivist death through sciences or its major death through auto-completion, through "thought" or "deconstruction." It is the problem of its global measure and determination through another experience of thought, which never had anything to do with it, and now less than ever. Any serious practice of philosophy is enough to show that the current North American slogans of "after-philosophy" and the "end-of-philosophy" have absolutely no conceivable sense: neither philosophical (especially when they are attributed to "deconstruction" . . .) nor even scientific. There is no effective end to philosophy. Neither can philosophy program its own destruction without reaffirming itself, nor can science envisage anything of the kind, since its goal is at most to rigorously know philosophy, but certainly not to replace it. Only positivist philosophical decisions that lack self-awareness can share, for a moment, the illusion of an effective suppression of philosophy. But if these decisions have some significance for thought, they have none for the real. Instead, with regard to philosophy's *claim* to provide the essence of thought, to legislate over science itself, we can say at once that it continues as before with the effectivity peculiar to it and, *moreover*, that it is already invalidated, displaced outside itself, *restored to its place outside of and by science.*

Our goal is to elucidate the new paradigm as well as to demonstrate the well-known insufficiency of every philosophy (not only of "metaphysics") to *sufficiently* fulfill the tasks of a simple, slightly rigorous description of scientific, artistic, technological . . . work. We are thus not concerned with the "postmodernist" or "deconstructivist" rescue attempts. It is always possible to believe that the work of sciences and of arts has been "covered" with the old apparatuses of philosophy, once they are brought up to date— philosophers are among the last handymen, those who do not yet know that their tool has been irremediably deficient since its birth and was cobbled together—but it is increasingly less possible to attribute the enterprise's inefficacy to the "height" and dignity of "thought" . . .

FROM THE IDENTITY OF SCIENCES TO THE SCIENCE OF IDENTITIES

Difference and *Identity* serve here as our guiding motifs for a reflection on the ultimate foundations of science and of philosophy and on the possibility of a scientific reform of the philosophical understanding. At first glance, Difference and Identity do not introduce us to science, but merely to philosophy. Yet the destiny of these "categories" is tied to the continuous expansion of their signification within philosophy, an expansion we now try to shift toward science itself and its ultimate foundations. *Difference* and *Identity* were at first simple predicables in a list of notions with reflexive yet secondary functions (cf. Porphyry, *Isagoge*). At the very most, we can admit that Identity communicated implicitly with the thematic of the One.

But this thematic, as we know, is subject to the greatest uncertainties and to illusions whose solution—precisely with respect to Identity and its relation to Difference—would have to touch on the future of philosophy. With Kant, they acquired the superior, always "reflexive" status of concepts that regulate the "amphibological" use of the understanding and that bear on the whole of philosophy. With contemporary thought since Nietzsche and Heidegger, Differe(a)nce and in turn Identity acquired a still higher dimension as "transcendentals" of a new kind, as principial [*principielle*] categories. Identity is capable of summing up the entire metaphysical or representative thought, differe(a)nce the thought that seeks to critique, delimit, or deconstruct it.

But in all these cases Differe(a)nce and the correlative Identity in no way express *a theory of philosophy as such.*[1] They express the perspective of a singular philosophy, a gaze on the tradition as a whole, which nevertheless remains trapped within it. We will attempt to impose an ultimate extension—in the form of a scientific thought *of* philosophy rather than a new philosophy—on these "categories" and thereby to destroy their uses as "overcategories" of the old ontology or as "transcendentals" of the deconstructions of this ontology. Difference must be able to describe (without any remainder) philosophy in its most enveloping concept, and Identity the theory of philosophy, i.e., a science that can problematize philosophy and that has *at least* its "power."

We note that this new use—first for science and its theory, then for the relation of science to philosophy—implies a "reversal" of the priority between the two. In philosophy and on its margins, Difference overrides Identity. This is even how we will define philosophy, in its telos at least: by the primacy of Difference over Identity and by what this primacy implies, their system or their correlation, one being inseparable from the other, the two reciprocally determining each other, Difference itself presupposing Identity. On the other hand, in the new theory of science, which can be established by treating these notions, and especially Identity, as a guiding thread, Identity overrides Difference (i.e., philosophy), although now it is no longer a matter of domination or primacy but of simple priority or order. It is still necessary—we will undertake this—to emancipate Identity from Difference's traditional authority, to emancipate it more exactly from their *system* or their *correlation*, which sustains the primacy of Difference. We must be able to render them heterogeneous to each other, in such a way, nonetheless, that their system (philosophy itself) is not simply fractured or deconstructed (according to Difference), but no longer serves as the foundation of their relation. It is no longer but a "term"—Difference itself—coming after Identity and its priority. The new problem is thus the following: *how to think Identity outside-Difference, i.e., as such?* How to rectify its concepts and its ontological statements so as to render "thinkable in itself" and plausible the existence of such a thought, which no longer moves within Difference—within the *Two* as much as within the *One*—but within and through the *One* alone, maintaining with the Two a new and freer relation? If Difference and Identity are inseparable in philosophy, reciprocally determinable from the standpoint of their knowledge and their reality, they are so in science only unilaterally and only from the standpoint of Identity's *representation*, not from the standpoint of its reality. *Difference is necessary only to the knowledge of Identity and not to its essence. This is to say that Identity and knowledge, the real and science, are not reciprocal or convertible.* What destroys convertibility is "determination-in-the-last-instance."

The most banal slogans, the most basic statements like "the identity of a people," of an individual, or of a phenomenon, are this science's guiding formulas. Guiding because they are symptomatic or indicative of a problem: do identity and its object, that in which language says identity, co-belong or not? Do they reciprocally determine each other or not? A science of Identities sets out from the dissolution (without a protocol, absolutely given) of this bond between language and representation—philosophical representation or indeterminate generality in which, on one hand, Identity and, on the other hand, Being, being, or the Other are supposed to convert into one another, to co-belong and to form more or less reversible blends. It is not enough to suspend every structure of intentionality and of ekstasis ("identity-*of* . . . ," in the style of "consciousness-*of* . . . "), if this suspension aims to maintain the simultaneity of the One and Being. We have to start by recognizing the autonomy of the One as real *before* Being and the Other, therefore before singularities themselves. Identity-of-the-last-instance opens then a new field of descriptive and theoretical possibilities. But on the condition that it is no longer considered directly legible on the surface of objects or logical forms, of Being or the Other, any more than concealed behind them and "withdrawn" from the World. Identity-of-the-last-instance is independent of that of which it *seems* to be said. It is a *real* essence, an essence that has never been an attribute, and, for this reason, since it is the Given anterior to every Donation, it does not constitute a back-ground like Being, for example. It is the World that is ultra-One; it is Being that it trans-unary.

If this problem can be resolved in all theoretical rigor and without the help of philosophical methods—but perhaps with the (henceforth secondary) help of philosophical statements—then it is indeed the future of thought before or after philosophy—outside it at any rate, but in a new relation to it—that will be implicated in this new bifurcation, which is no longer between Difference *or* Identity, but between the *domination* of Difference over Identity *or* the *order* that goes from Identity *to* Difference.

To this end, it is enough to show that the One *qua One* or qua emancipated from Being is the object of a science rather than of a philosophy, but also that this science of the One or of Identities allows us to describe the essence of sciences, the sense of scientific labor, in a nonepistemological description (or a description emancipated in its turn from philosophy). It is in this way that science (as nonepistemological) and the real or Identity (as nonontological) intersect: both are nonphilosophical. This task is accomplished in a new *theory of science*. The interest of this theory—we obviously do not describe it here—is that it combines, to the point of radically identifying them, the *essence* of Identity and the *essence* of science. On one hand, it affords science a real or transcendental dimension it lacks in epistemologies, which implies that science is not an abstraction, a formalism, a dialectic . . . but is directly related to the real itself, as philosophy claims to be, *with* the difference that it is actually related to the real. On the other hand, or in parallel, this real to which science relates its knowledges is no longer Being, but the One, which stands outside-Being, outside-presence or outside-representation: in itself alone, or, as we will say, it forms a "determination-in-the-last-instance." In other words, it is inalienable in the theoretical representation that makes it possible and that is nevertheless *its* representation or the knowledge that determines it. Hence the double exclusion to which we referred and that is explained by this "real" or "transcendental," nonempty character of science, which, in the form of the science of the One or of theory of Identities, can treat philosophy, its sense and its language as its objects, and can do so without contradiction. There is an affinity, even an identity, between the essence of every reality, the One-of-the-last-instance, and the essence of science. Only philosophy can desire to separate them, to separate science from what it can do and to refuse to acknowledge the existence of this original thought.

The elaboration of this new science—the science of Identity qua Identity and thereby the science of the *essence* of every science—amounts to formulating a kind of "general method" for posing and resolving problems. It calls science and philosophy to the same task, but in new relations. "General" no longer has here the sense of *"unitary* theory" (primacy, domination: of philosophy over science, of Difference over Identity), but the sense of *"unified* theory" (simple priority, without primacy, of science over philosophy, which is treated as a specific "object" of this science). It is a matter of extricating thought from the state of separation with what it can do, to which philosophy's domination consigned it, and reforming the understanding in the nonpositivist direction of a scientific *thought* that continues to make use of philosophy, of philosophical statements about Being, although it is nevertheless not a philosophy and above all not a "philosophy-as-rigorous-science." It is science that comes from the depths of itself "to" philosophy: *to meet* it. Not in the sense that it starts to philosophize and loses itself in the process, but that it institutes itself as nothing-but-science and thus maintains a new relation to philosophy.

We describe here and there some aspects of this general method, articulated on the priority of Identity, insisting above all on the new freedoms it gives thought. Precisely because it is massively a thought of and by Difference, philosophy treats nothing but *amphibologies.* This is what will in turn cause the science that mainly treats *identities* to appear. But it is a science of a new type: not extracted from philosophical mixtures, but globally "anterior" to them. Philosophy is the thought by and for blends (of orders, spheres, or instances). And it is not only the forms of classical representation that constitute an *economy* of differences and identities. On the other hand, the thought (of) science, thought-science, is the chance of an access to the *real*, to the *identity* of phenomena. There is little doubt now that this identity is no longer metaphysical or generally philosophical. We will say that it is *the Identity-of-the-last-instance of phenomena.* Any forgotten phenomenon of philosophy and often of already constituted sciences, from the most concrete (the book, photography, etc.) to the most abstract (the real, the ideal, the understanding, intuition, judgment, etc.), can now become the object of a science that relates it in a specific way—for which science alone is the key—to its identity. Science rediscovers the orders, spheres, or instances against philosophy, which, blending them systematically, can think none of them, but only "thinks" itself.

Under the name *Theory of Identity*, we propose to establish the principles and most of the techniques of this general method for the description of any phenomenon whatsoever (phenomenology should have been this method, but it failed because it exempted itself de facto from the description). This discipline does not claim to replace existing sciences, but rather to replace philosophy, which always excepts itself, not so much de jure as de facto, from its own *theory* and compensates for this failure of its theoretical project with the illusion of its hermeneutic omnipotence. All phenomena no doubt—*but only from the perspective of their Identity-of-the-last-instance*, and from the standpoint of the Identity through which they cease to simply belong to the World or to philosophy—can enter into the vision not of an already existing science (they are already the object of various sciences), but of this science in which we demonstrate and know what it means for them to be objects of science and to cease belonging as such to the World or to leave philosophy's sphere of influence.

Unlike other rival theories (catastrophes, multiplicities, critical points, language games, complexity, self-organization, etc.), the theory of Identities is not obtained through a reflection on current scientific knowledges, isolated from their context; on particularly seductive or promising theoretical givens; or on longstanding theories, accepted as a kind of scientific common sense (axiomatic set theory, for example) and abstracted from the immanent scientific process. It is not a *philosophical artifact* engendered by the present state of sciences and doomed to imminent obsolescence, like the theories that only hold up because of philosophical inertia rather than scientific movement. It is elaborated by a science of the *essence* of science, of its status vis-à-vis philosophy (even when the theory of fractals is at issue), a theory of what it *can do* as science and so forth. Far from representing a simple effect produced by the impact of a determined theory on a fluid, indeterminate philosophy that is ready to immediately reorganize or rebalance itself, it represents an attempt at a global reevaluation of the relations between science and philosophy, a redistribution of their exchanges. A redistribution that is simply the description of their real distribution. This description does not merge with their philosophical image, which forms a system with the continuous production of ideological, half-scientific, half-philosophical artifacts, whose mixed nature ordains them in a privileged way to mediatic consumption. It originates in the "spirit" of science alone, in the *thought* of which science is capable and that has no need of philosophical support.

The denounced artifacts are scientific yet local knowledges, abstracted from the process of scientific thought or from their scientific sense and associated with another type of thought (the philosophical) in order to compensate for their initial abstraction. They result from a dismemberment of theories. Philosophy cuts off theory from what it can do; it isolates the knowledges produced by science from their immanent theoretical-being and then substitutes itself for science, claiming to stand in for it or to fill in what it believes to be a void or an absence. It thus obliterates the *identity* of knowledge and of thought-theory; it appropriates the former and negates the latter. It then produces these strange beings, these half-scientific, half-philosophical chimeras: "multiplicities," "catastrophes," "singularities," "inconsistent multiples," and so forth.

A theory of Identities does not rest on this ground and rejects these traditional customs of philosophy. It merges with the autodescription of thought-science, of theoretical-being and of its cause. It positions itself precisely at the point of the Identity-of-the-last-instance of the real, of knowledge and of thought. A particularly ungraspable, opaque point, so barely illuminated in itself by the light of logos that this light makes it vanish without delay. Hence the indefinitely varied distinctions between knowledge and thought, theory and philosophy, metaphysics and thought—so varied that the efforts to surmount them only revive or reproduce them in the most *symptomatic* fashion, in the form of artifacts that represent nothing but themselves, the unreality of philosophy's suture to science. Identities, as they are immanently postulated by science's theoretical practice, are more primitive, simpler and more originary, if possible, than their philosophical equivalents, which are transcendent constructions, always labored and unstable, endangered by the elaboration of knowledges and the falsification of theories.

A theory of Identities has therefore a scientific and nonphilosophical origin. It rejects all these "negative" and transcendent traits of singularities. Instead of restarting with singularities as a "new" beginning, it starts out with the beginning, with Identity as such. Far from being a last mode of ontology's autodissolution, it is founded on the immanence of *real* Identity (to) itself, on its absolute nonconvertibility with Being (philosophy of the Ancients) and with the Other (philosophy of the Contemporaries). Conceived in this sense, Identities do not sacrifice singularities, but only their exteriority and their folded nature as mixture or doublet, their transcendence in relation to the real, and, lastly, their generality and indetermination.

Identities' radical description, that of their irrevocable precession on every representation (even the theoretical), gives a new impetus—at the same time that it alters their nature—to contemporary questions about philosophy, to its association with an Other-than-philosophy. Philosophy itself has only been able to contest its own claims (in a limited way) through practices of the relation to the Outside, the Other, the Margin, the Nonphilosophical, and so forth, insofar as they are "without-relation." But it continues to presuppose the whole of its right, the essence of its validity, and admits, in the same style, an "Other" or an "Outside" it cannot found. This double presupposition breathes life into philosophy, but is *theoretically* arbitrary and cannot be accepted within a theory of Identities. At most, it can become an object, a "phenomenal given," of this science that hence *radicalizes* the critique of philosophy, in other words, alters its nature. A critique that ceases to be entrusted to philosophy itself in order to be entrusted to a science.

FRACTALITY AND CHAOS: THE OBJECTIVES OF A GENERALIZED FRACTAL DESCRIPTION

The restoration of Identity's essence (and thus of science's) is the first step toward a more rigorous position of the problem of philosophy. Before reaching this threshold, we will cross a second stage that will form the transition between the science of singularities and the science of philosophy: the elaboration of the concepts of *generalized fractality and chaos*. Since we are prohibited by definition from proposing a *philosophy of fractals*, which would once again interiorize the geometric and physical fractals (Benoit B. Mandelbrot) into the concept and through the system of philosophical operations, we will use them as a simple indicative material in elaborating the *generalized concept of fractality*. This concept will rest on a mutation in the theoretical basis of fractality. And it will place us directly on the path of an *Artificial Philosophy*: a science of philosophy that proceeds by the "fractal" and nonphilosophical synthesis of statements.

The geometric theory of fractals belongs to the set of new objects (catastrophes, singularities, differences, games, turns, etc.) that contemporary thought has discovered and exploited against metaphysics. Yet it stands out for its scientific status, and we treat it here as our guide. It has undergone theoretical extensions (from one region of physics to the other or from physics to economics, etc.), but also aesthetic and sociological "interpretations." The extensions and the interpretations do not respond to the same criteria and do not exhibit the same type of rigor. We propose to rework them here and to afford them more than a regional theoretical extension and more than a free and metaphoric interpretation or an aesthetic transfer, although it is for us a matter of realizing the identity of this double objective, rendering these two uses identical. This is what we call *a theoretical recasting of fractality, a change in the theoretical element of fractal objects*. It is a matter of elaborating the concept of a *generalized fractal* that could be quickly called (subject to precisions and nuances) non-Mandelbrotian. To be sure, this enterprise requires extremely precise conditions so that it does not founder in a simple "hermeneutic" requisition or a groundless transfer of Mandelbrot's discovery. To give an initial idea of the difficulty in elaborating this theory:

- it must be scientific, yet not geometric or a physical application of geometry; theoretical, yet not necessarily mathematizable;
- it must be formulable in natural language and able, moreover, to define a fractality that affects this language, yet it must not remain philosophical.

This double exclusion—of geometry, but not of every science; of philosophy, but not of language—negatively defines the field to be explored. It is, in some sense, the "negative" of this undertaking. The enterprise obviously emerges as a paradox in the eyes of scientific common sense (for which there is no science other than mathematizable science) and of philosophy (for which there is no "possible" science of language and sense other than philosophy itself). The paradox of this type of generalization of fractals is clearly the same as the paradox outlined by one of our guiding formulas, one of the possible objectives of this research, the Idea of a *science of philosophy*. It does not go beyond this objective and is even inscribed in the program of a science or a theory in the strong sense. But a science that is not mathematizable (thus unacceptable according to the "spontaneous" ethos of the scientist) and that bears on an object of language and sense that has, up to now, always resisted its reduction to the state of a science's object (thus unacceptable according to the ethos of the philosopher). Here the exclusion is fertile insofar as it is double and bears simultaneously on the geometric version of fractals and on their eventual philosophical version (whether it already exists or not is of little importance, it is entirely imaginable). But how is it possible, what concept of science, of a science that does not yet exist (that of the *essence* of science, thus of philosophy) must we develop—and could we?—to render credible or simply plausible this scientific break in the theory of fractals, this *generalization*, which can only be scientific and no longer philosophical?

In reality, we define the question in this way only because we already have at our disposal, not the formal and elaborated solution, but the problem's givens and thus the possibility of the solution. This generalization of fractals, which is neither a "regional" extension nor an interiorization and an ideological capture, is one of the unexpected consequences of the nonepistemological theory of science. It founds itself on an older discovery: the "grounding" element or the cause of science is not *Being*; rather, the *reality* scientists speak of is much closer, albeit not identical, to what philosophers call the One. Provided we first emancipate the One or Identity from their philosophical use, from their forced companionship with Being. This is not yet the appropriate place to elaborate such a theory of science; we did so elsewhere (*En tant qu'Un*) and do so *right here* in the first chapter. But this sketch is enough to show that the paradox of a science of philosophy, and thus the paradox of *a generalized fractality under the conditions of this science as this theory formulates it, and which are therefore neither mathematical nor philosophical conditions*—that this paradox only exists in philosophy's eyes. The path is open for the recasting of fractality. It is hardly necessary to say that this recasting contains no critique (philosophical or mathematical) of Mandelbrot's fractality, whose generalization under these precise conditions is by no means its negation. It is not a matter of a rival theory, but of the only theory capable of safeguarding geometric fractals from their interiorization and philosophical capture, from what can only dull their virulence.

A generalized fractal description responds, then, to several interconnected objectives:

• The first is to extend the geometric concepts of *dimension* and of *fractality*, first from objects to science, to scientific thought itself, then to philosophy. Put differently: from nature to thought, from so-called natural and regional phenomena (turbulences, economy, meteorology, theory of aggregates and interfaces, etc.) to those of the representation of these phenomena, to the absolutely universal sphere of knowing. This generalization is thus not "ontic" from one region to the other; nor even "axiomatic" (a "non-Mandelbrotian" fractality, although we will re-use this term); but from being [*l'étant*] to Identity, to science as the thought of the One. A transcendental yet nonphilosophical extension, required by science itself, which uncovers its authentic sense. If so-called natural properties are merely reified knowledges, it becomes possible to carry out this extension from "ontic sciences" to "first science." In this way, the fractal modeling of philosophy itself—instead of a philosophy of fractals—becomes possible in the form of a "philosophy-of-synthesis" or an "artificial philosophy," which is nothing other than "nonphilosophy" (in the sense in which this word designates the final product of our undertaking). We also have to first remodel and generalize the concepts of fractality and chaos under nongeometric conditions, the conditions of the essence of science itself, without thereby producing a philosophical generalization that could only subdue "scientific" fractality once again.

• The second objective is to detect, to give consistency and above all a clear formulation, to a new problem overlooked by philosophies, whether they are philosophies "of" sciences or . . . "of" philosophy itself: the problem *of a definition and a measure of the reality-content of knowing.* Here it is a matter of a new property or determination of knowings, of knowledges or of thoughts, which would have to liberate them from philosophical authority. If philosophy merely examines knowings themselves, the idealized knowledge-contents it appropriates, *what is at stake then, with the problem of the content-in-reality of these knowings, is an absolutely new problem, a theoretical discovery,* which alone suffices to justify this research that goes from Identities to Artificial Philosophy. This is a discovery as crucial for the non-epistemological theory of science and of philosophy as the discovery of the content-in-fractality of phenomena that have seemed, up to now, to pertain exclusively to continuous physico-chemical or geometric processes. It is a matter of a "structural" property that touches on science in its essence and thereby on philosophy, but which cannot be explained by continuous processes like those that philosophy and its epistemological images can exploit. It is the problem of the *dimensional* nature of knowing in general, and in particular its *dimension of reality*, which cannot be reduced to the dimensions of *possibility* and of *effectivity*. It is a matter of measuring, qualitatively, of course, this property—new in relation to their idealized knowing-contents—of philosophical decisions and bodies of knowledge. This property is their fractal dimension. We will show that it maintains the closest links to real Identity. The dimensional description of philosophy, of ontic sciences and of first science is what allows us to distinguish them, to extricate them at last from their unending philosophical confusion and to inaugurate new, more fertile intersections. Among these dimensions, the fractal will be the most complete and the most decisive; it will allow us to measure the qualitative degree of reality of the figures of thought, for example: "singularities," "multiplicities," "differences," "language games," etc.

• The third objective is to make use of this generalization of fractality as the long-awaited solution to an old, unresolved problem: the possibility of an "artificial Philosophy" whose concept is not uniquely informatic and restricted to this technology.[2] Even though the principle of the solution has been discovered with the scientific position of the very *problem* of artificial philosophy, the theoretical tool of generalized fractality will be necessary to make the idea of such a *synthesis of philosophical statements* plausible and noncontradictory. As generalized, fractality has to allow us to pass from the aporias of an artificial philosophy via informatics—which would now represent only a local means—to the concept of *statements of synthesis* that rely on fractal procedures. We must not forget, in fact, that the main upshot of this generalization is the extension of images-of-synthesis to phenomena of natural language and sense, the extension of sense and of thought itself.

Once this theoretical threshold—a fractality of identity and no longer of difference—is crossed and once the unraveling or unwinding of philosophical resistance is realized, the formula "science of philosophy" returns to the scene and takes on a new sense. Not only is this guiding formula intrinsically grounded and legitimated through the idea of a generalized fractality, but it also develops concretely as a fractal theory *of* philosophy itself. It is science or theory that *is* fractal and that fractalizes. On its own, philosophy is not and refuses to be fractal. Instead, it represents the condition of continuity, the continuous curve that tolerates a minimum of—if any—fractality. It can and must be fractalized in its turn. And in a way all the less external and "artificial" because it is par excellence the object fractalized by the science of the One, the support or material that receives the fractal structure. Philosophy is one of the *internal* conditions of this structure, but in such a way that it does not constitute it and that this structure comes to it from the outside.

We establish in this way the theoretical basis for a fractal description or modeling of philosophy. The theoretical, yet transcendental and nonmathematical generalization was needed in order to extend this theoretical tool to philosophy, a grounded extension despite the resistance of philosophy, which rejects this new science and tries to interiorize it anew. In its geometric state, fractality is hardly useful for phenomena of *sense* or of *philosophy*, i.e., of *language-in-philosophy*, of *logos*. It is probably useful for linguistic phenomena, signifier and signification, but not for *logos*. We now have the theoretical bases that allow us to potentialize it and to afford it the transcendental dimension that its new object (philosophy) requires. The bases to force philosophy to cease capturing singularities, critical points, catastrophes, and scientific fractals, and instead to *fractalize logos itself* in a radically heteronomous way (this heteronomy is fractality itself, its transcendental and nonempirical or given-in-*the-World* sense).

But this triple objective refers to practice. The preliminary task will be to found the new concept of fractality on a renewed theory of Identities; this concept will be valid for natural language and no longer uniquely for physico-geometric nature. Philosophy has two complementary concepts of Identity at its disposal, which it divides as follows: *Principle of Identity* (logical) and the *Same* (logico-real), *plus* all their intermediary modes (unity, subject, bond, synthesis, difference, etc.). Neither concept can constitute the terrain of this fractality; it must be real and not simply a possibility dissolved for the benefit of philosophical generalities, which are partially indeterminate and "efface" fractal rigor, which cannot even grant a right, an *unconditional* right, to geometric fractality, a fortiori, to the fractality that can be generalized to scientific thought itself or as such. The theoretical elaboration of a radical concept of Identities corresponds to the opening of a new space of thought. It is in this space that fractality can be reformulated as an "internal" or "transcendental" property of knowing itself and not only of some of its objects; it is in this space that it finally allows us to resolve the otherwise impossible problem of an artificial philosophy.

The first and second parts of this research supply what should be termed the "conceptual" (rather than mathematical) "formalisms" necessary for the establishment of a science of philosophy. They make what philosophy would have called "concepts" or "categories" manipulable; they transform them into exportable tools, which can be transferred to other regions of knowing. This is not to say that the last part is the most important and that this essay "culminates" or "concludes" in a theory of the philosophical decision. Scientific knowing does not recognize such philosophical hierarchies and teleologies, and the theory of Identities is as crucial as that of fractal Identities and that of philosophy. Because it is a question of the same "irregularity," the same "fractality" from one end of these inquiries to the other. Their "self-similarity" . . .

PART I

THE ESSENCE OF SCIENCE

1

SCIENCE

A NONEPISTEMOLOGICAL DESCRIPTION

SCIENCE'S ANTE-EPISTEMOLOGICAL RECONSTRUCTION

THE possibility of the "philosophy of sciences" and of "epistemologies" is an evidence that philosophers rarely interrogate. We interrogate it here by means of science itself, of a nonepistemological description of sciences. Is it certain that philosophy can legislate on science, found *and even describe it?* . . .

One way of renewing a phenomenon's interpretation is to inventory the points that traditional interpretations systematically forgot or deemed uninterpretable, to coordinate them and to found on them a new global theory of the phenomenon such that it is not opposed to precedents, but subsumes them as particular cases. This is what we aim to do here in the case of science and with what philosophy and epistemology dismissed from science as "impossibilities." What was at once interiorized and excluded from science, interpreted and furthermore declared beyond every possible signification, is what we can call a basic or postural realism. Not the realism that science can sometimes exhibit through certain knowledges (from this point of view, it tends instead to dissolve every transcendent reality), but a fundamental or postural realism, with the phenomena that accompany and express it: 1. an opacity of thought such that it leads us to think there is no scientific *thought*, but a simple, symbolic manipulation without sense and reflection; 2. an immanence of theorico-experimental criteria; 3. a realism that seems to be that of the brute and transcendent being beyond every objective genesis of object-sense, etc. This realism seems thoroughly foreign to philosophy, which must then engage in an operation of division or distinction between what, from scientific practice, is capable of being founded by logos and what "falls" outside reason and its light (as an incomprehensible and almost irrational realism).

What can we do with this realist *claim* that emerges beneath every local knowledge? The first rule is to take the measure of its genuine radicality vis-à-vis philosophical thought and operations. The second is to thoroughly alter the hypothesis of interpretation and to believe that this immanent realism is the very Identity of science, its "cause-of-the-last-instance"; that it is *its own rule of interpretation or its criterion and that it "displaces" the ontologico-epistemological interpretation* in general and serves as its real foundation. Science is perhaps the great unknown of occidental thought. Not that philosophy is not devoted to it in the form of epistemologies and of ontological foundations. Quite the contrary: it is precisely because we have always had a philosophical and Greek vision of science that it remains unknown in its essence. Our experience of science is simultaneously marked by an underestimation (science as deprived of sense and of absolute truth) and an overestimation (science as factuality and effectivity) that characterize its philosophical and "cultural" interpretation. From Plato to Kant and Heidegger, a triple *division of intellectual labor* has reigned: 1. One admits that science produces knowledges, but denies that it thinks. To science, knowledges without thought; to philosophy, authentic thought, the one that necessarily needs knowledges, but that, on the other hand, founds them, legitimates them, and simultaneously supplies their genealogy and their critique. "Science does not think"; it "dreams," it only dreams thought in the very operation of knowledge. 2. There is an absolute, unique, and self-founded science—first philosophy as ontology or logic—and empirical sciences, which are multiple and contingent through their object; they produce strictly relative knowledges. Philosophy divides the concept of science after having separated knowledge and thought. 3. To philosophy, Being [*Etre*] or the authentic and total real; to science, not even being [*étant*], but the properties of being or the facts; the object of knowledge is now what is divided.

No philosophy really escapes this triple division of intellectual labor. No epistemology (whether empiricist or idealist, positivist or materialist) can break free from what is a rarely recognized invariant of science's Greco-philosophical interpretation, including the Anglo-Saxon one.

We propose to disabuse ourselves, once and for all, of this falsifying image that philosophy has always imposed on science and from which it extracts for itself a surplus value of truth, authority, and dignity. On the essence of science and its relation to the real, we maintain an altogether different thesis that requires us to do away with this philosophical appropriation. Against this triple division, we propose to admit *the Identity-of-the-last-instance of scientific thought* as hypothesis or axiom—an entirely different problem from that of the unity of sciences. By what method? That of an *immanent autodescription*, for science is capable of rigorously describing itself when it takes its radical or postural realism (i.e., this Identity) as an immanent guide to its own understanding. Rather than "reflecting" on "scientific knowledges" in view of an "epistemology"—a procedure that endorses the philosophical prejudices and fetishizing interpretations of these knowledges—we reconstruct scientific thought and its categories (hypothesis, axiom, theory, experiment, etc.) within the limits of their immanent description. *A reconstruction*: in order to combat its protracted dissolution, its interminable deconstruction by philosophy. *Of scientific thought as such*: insofar as it constitutes an autonomous essence, irreducible to the philosophical schemas that are founded on the philosophical Decision or the operation of Transcendence: logos. This thought of science, the thought of the thought proper to science, is no longer an epistemology but an *epistemic*: a theory of the *episteme* insofar as, though named in Greek, it has no need for Greco-philosophical logos. *Within the limits of its immanent description*: we situate ourselves within the essence of its practice in view of describing it in its immanence, bracketing through a radical reduction—more radical than Edmund Husserl's (we will come back to this point)—not only some philosophies like empiricism and idealism, but every possible philosophical position in general, i.e., every operation of Decision or Transcendence. In this way, we regrasp and elucidate the *immanent phenomenal givens* that constitute science's essence, guarantee it an autonomous and specific reality, and avert its philosophical dissolution. We describe its essence before it is divided by philosophy into opposites, in terms of adverse positions that struggle against each other in order to impose a meaning on science and draw a surplus value from it: empiricism and rationalism; materialism and idealism; anarchism and dogmatism, etc. This description will be *transcendental*, but neither in the

Kantian nor the Husserlian sense, because it now coincides with a real "last instance." Only a transcendental, i.e., rigorously immanent description that has ascertained the means of avoiding any constitutive usage of metaphysical transcendence (philosophy in the Greco-traditional sense) can "found" not only the objectivity or the possibility but also the reality of knowledge against its empiricist factualization, its hermeneutic dissolution, its idealist sublation. This transcendental "determination" of a new style amounts to grasping, describing, and legitimating science's *Identity-of-the-last-instance* and to distinguishing it from its philosophical masks.

Through this description, carried out in a spirit of submission to the *requirements* of scientific thought, we hope to render service to scientists rather than to dispossess philosophers. This conception radicalizes certain positions of Husserl or Kant and draws out all their consequences: they are not *anti*philosophical or *anti*epistemological, but *ante*philosophical. We extract a real, *ante*-epistemological ground from which it then becomes possible to reevaluate, in their entirety, the status and the function of epistemologies and their role in scientific labor itself.

CRITIQUE OF THE PHILOSOPHICAL DIVISION OF SCIENCE

The majority of philosophies that sought to interpret science presupposed implicitly, sometimes explicitly, that it first presents itself in relation to them as a sort of radical alterity, a reality beyond even the reality of the object as this latter can be engendered within a genealogy of objectivity, i.e., of the object-sense. Philosophy can only recognize a certain autonomy to science on the condition that it treats science as its Other (an Other as philosophy can conceive it), that it sunders it a priori between, on one hand, a *factual* brute and transcendent existence, a token of the autonomy it must recognize in science, as well as a sign of its basic opacity and unintelligibility, and, on the other hand, an objectivity it can always engender in its sense and ground in the essence of the philosophical Decision. This Decision, failing to wield a power of legislation over the most transcendent real, would have in any case such a power over the *relation* of science to the real.

Such is the invariant of epistemologies and of ontologies: the interpretation of scientific realism proceeds by *dividing* it and thus dismissing

science as philosophy's Other. That science can only acquire its autonomy by stressing the aspect through which it is philosophy's unthinkable
Other is obviously already a philosophical perception and a philosophical
falsification of the reality of science. That it can only be thought by
philosophy—whatever this latter may be—at the price of its *division* and
of the destruction of its autonomy, a division that is in fact the nucleus
of the philosophical Decision, makes us realize that we have to alter the
general hypothesis of interpretation, but it does not force us to do so.
No doubt it is no longer a matter of exchanging the idealist-critical hypothesis for the positivist, the empiricist, or the "realist" (in the limited sense
in which philosophy, analytical or not, can be realist). It is philosophy in
general, insofar as it always includes—this is an invariant—a moment of
transcendence (scission, nothing, nihilation, difference), *which is deemed to
be constitutive of the real*, that we have to put once and for all out of play as
soon as it is a matter of science, its autonomy and its essence. An attitude
of descriptive fidelity, phenomenological in a radical sense since it allows
the phenomenal givens of science to describe themselves immanently: such
is the only "method" adapted to this philosophically impossible object; the
only one that establishes the autonomy of science's thought in terms of its
proper description and therefore implies the suspension of epistemological
positions and operations. It suspends from the start the thesis that science
is an Other or even the Other of the philosophical Decision; this thesis
already contains a priori the necessary *division* of its Identity and produces
incalculable effects. The division of its essence alienates or tries to alienate
science outside itself. It idealizes or derealizes science like every bar that
strikes through a subject; it renders it transcendent and ultimately unthinkable; it deems it to be unthinking and deprived of reflection.

Our hypothesis treats science's postural or "subjective" realism as a transcendental guide or rule of its immanent theory. It is no longer, we suspect, a
hypothesis in the sense of a *possibility*—a simple philosophical *possibilization*
of science—which would obliterate its reality by subjecting it to the play of
the possible and of the hypothetical (philosophy reserves the anhypothetical
for itself). It is the scientific attitude that consists in entrusting to science
itself the elucidation of its essence, in recognizing its radicality to the very
end, and in drawing all the consequences of its autonomy. Science is for itself,
at least in its cause (the Identity-of-the-last-instance), an emergent theoretical object, a "hypothesis" or an "axiom" in the "hypothetico-deductive"

sense. We also have to discard once and for all the idea that science is a subontology, a deficient or "regional" ontology; a "rational fact" to be interpreted; or else a simple process of production of knowledges that ontologos or epistemolgos could delimit and protect. In general: the already philosophical presupposition, which would be the Other of the supposedly first or valid philosophy. This "change in the terrain" restores science in its proper reality or reconstitutes it as *its own immanent basis.*

Science becomes "first" and thereby ceases to be philosophy's Other. It is philosophy that becomes science's Other, but the "Other" in a non-philosophical sense, for it is determined by science. The global reevaluation of science's essence, in which we must engage, is the necessary and sufficient condition for a new position of its relations to philosophy as well as for the elaboration of a science of the philosophical Decision.

THE TASKS OF A RECONSTRUCTION
OF SCIENTIFIC THOUGHT

If we wish to guarantee science its full autonomy vis-à-vis philosophy, three simultaneous tasks await us. They correspond to the critique of the division of intellectual labor and amount to showing that every science ("empirical" or otherwise) is equally a thought; that it is absolute *in its kind*; that it bears—at least "in-the-last-instance"—on the real "itself." Here is what we will have to demonstrate:

1. Science does not confine itself to knowing; there is an authentic *scientific thought*, i.e., a relation to the real "in itself." This relation is presupposed by the fact of knowledge and affords knowledge its *reality*. Science is an autonomous thought, altogether distinct from the philosophical. It does not use the philosophical thought's procedures of positional transcendence (decision, reflection, division, nothingness, nihilation, etc.). Science thinks without *founding itself* on transcendence in general; and when it finally resorts to transcendence, it is neither the philosophical that is at issue nor the operations that flow from it.

2. Every so-called empirical science, even if it lacks a mathematical dimension for instance, is absolute in its kind. It is guided by the Idea that

it is a science and a knowing (of) self in the mode of Identity-of-the-last-instance and not only a blind production of knowledges. Its theoretical criteria are immanent to it; it does not expect to receive them from philosophy.

3. Every science that philosophy calls "empirical" is related to a real-One, such that it can grasp this real-One beneath its ontological division between Being and the properties of being. Every science falls outside ontological Difference. This difference glosses over *being in itself* or the One by dividing it, or rather by covering it with the division into Being (or object) and into objective properties. This division makes the One vanish, and science along with it. In reality, science relates the knowledges it produces to an "objectivity" or an "exteriority," but this objectivity does not have the form of ideal objectivation of which philosophy is the theory and practice. Science has a *real object*, but this object derives from its cause or from the One, and it *is an object at the same time as One-in-the-last-instance.* Philosophy cannot grasp this thesis. It separates objectivity and the real, which it also calls "in itself." It then confounds the genuine real or the *real object* (in a more originary sense of the word, the one to which sciences are related) with the ideal objectivity and its operations (reflection, categorial objectivation, decision, transcendence, etc.). There is a quasi "ontology" of science, entirely different from philosophical ontology (whether it is substantialist or idealist or functionalist, little matters). We propose to excavate it through a rigorous description of its most immanent phenomenal givens. Why these?

The demonstration of this triple objective—science as *Identity* in which the triple division of labor is in fact impossible or is an illusion—effectively supposes that scientific *thought*, far from being reduced to knowledges that are related either to "facts," to a "categorial objectivity," or to "interpretations," is constituted (but in-the-last-instance alone) from *absolute immanent givens*, that it is thus a knowing of the real in itself and a knowing (of) this knowing, and that this existence of absolute immanent facts is compatible—against philosophy's thesis—with an objectivity or an exteriority, with a stability, and lastly with a reality (rather than a simple possibility), which are its most general postulates.

All this—the cause of sciences as "absolute" knowing (not in the metaphysical sense, but in the sense of the One-of-the-last-instance) and the theory of their object—remains to be demonstrated. Rather, it has to be shown or exhibited so that we can render sciences autonomous with respect to every philosophy. But it can only be demonstrated if the "method" is rigorously adequate to its object, if the science practiced here is more than a fact (empirical or rational), if it is a *real experience or a knowing (of) the "One."* The sole guiding thread of the description of scientific thought must be science itself in its immanent *reality* or the autonomy of its theoretical criteria, rather than alleged "scientific facts" or "knowledges" or even a "categorial objectivity." All of these are already transcendent, already philosophically interpreted phenomena, and they "obviate" the description in an "epistemological" sense. From the moment they are faithful to *epistemo-logos*, all epistemologies are interpretations that transcend science and are falsifications of its essence. The real must itself determine in-the-last-instance the method of its description. The absolute immanent givens that constitute, if not the whole, at least the cause of scientific knowing as *thought*, require a more-than-phenomenological or nothing-but-phenomenal description. Through their radical immanence, these givens are necessarily *phenomena-without-logos* and can now be described only through themselves, without the procedures of logos.

We suspect that this description, which is a science, bereft of the means of philosophical reflection and construction, will be fundamentally "obscure" or "blind." But it will not be for that reason—here lies the whole thesis of the *reality* of scientific thought—simply technical, manipulative, or symbolic.

Science's *quasi* ontology: this term does not designate the ontology that the great philosophical decisions of Idealism and Materialism, of Rationalism and Empiricism, presuppose. It designates the ontology presupposed by the scientific posture and its immanent claims on the real. These claims are completely foreign to the Greek ontological presuppositions, which serve as the horizon of epistemological decisions. Our method will be that of a radical phenomeno(logy), of a description of the most irreducible or indivisible givens. It will eliminate from the start their empiricist and idealist reductions, which deny them every authentic reality or ground them in philosophical objectivity. There are three types of these phenomenal givens. First, those of the *reality* of science (which is not dissolved in a simple *possibility* the way it inevitably is within philosophy), i.e., this Identity that serves as its autonomous and absolute cause. Then, those of its *objectivity* (the transcendence of the *real object* to which it relates its knowledges). Instead of founding science's reality on its objectivity, we found its objectivity on its reality. Lastly, those of its power of knowledge or of representation. They completely exceed the side of "knowledges," of which they guarantee the reality or the transcendental claim.

Thus the central problem of this quasi ontology of science is its *immanent phenomenal givens*. They constitute the science of the Identity (not philosophical, not circular or reciprocal, but in-the-last-instance) of a reality and of the knowledge of this reality. What is meant by the real-One to which science posturally refers and that would no longer be deducible from a categorial or ontological objectivation of the philosophical type, i.e., from transcendence in general? As a first indication, we can situate the proposed solution in the neighborhood of Husserl's solution: he too admits absolute phenomenal givens, but as a philosopher he continues to associate them with objectivation and representation (intentionality); he does not found the latter irreversibly on the former and does not modify their nature. Our thesis is that science, and science alone, realizes a posture that nearly all philosophers—Husserl especially—rejected as impossible: "radical transcendental realism." This realism is effectively inconceivable at the interior of the most general Greco-ontological presuppositions. They exclude this realism—without seeing that it is of-the-last-instance—on account of the idealizing or objectivating transcendence, and presuppose a circle, a reciprocal determination of the real to be known and of knowledge, of the *real* object and of the *object* of knowledge. Science's *postural realism* requires, on the contrary, the end of this circularity. On one hand, that the real or the real-object—henceforth invisible de jure in the horizon of philosophical objectivity—is an immanent given of-the-last-instance means that it cannot be transformed by the knowledge produced of it (do not conflate the observer's intervention in the production of the object-of-knowledge with his impossible intervention in the real object). On the other hand, its representation in the object-of-knowledge is descriptive and not constitutive. It does not claim to transform the real, but only transforms old knowledges and technico-experimental products. It does so by affording them a status as "simple" knowledges that, by their sense as "scientific" and not philosophical knowledges, confine themselves to describing the real object to which they refer in an immanent yet in-the-last-instance manner.

THE VISION-IN-ONE AS CRITERION OF SCIENCES

To ground the real's scientific phenomenalization in its reality, rather than in its simple "real possibility," in the way we have just assumed or required it, and to distinguish it permanently from that of philosophy, we will ask whether we possess, and in what concrete way, a sufficient experience of it, an experience that can give science its autonomy of thought and thereby—this is a consequence of the structure of its essence—its greatest primitivity, its anteriority to the philosophical Decision.

How is the internal architecture of this special "transcendental realism"—which ultimately renders it noncontradictory—possible? What do we call a *radical immanent given* (to which science is not reduced, but which is the cause of its specific objectivity)? If the essence of the real to which science postur-ally refers excludes, in its intimate constitution, objectivity and all the forms of philosophical or positional transcendence (dialectical scission, alienation of consciousness, categorial representation, transcendence of the ontological project, intentionality), this is because its *immanence* must have an original nature. Whereas philosophy regularly combines immanence with transcen-dence, Identity with Scission or Difference, etc., so as to strengthen the one by the other, so as to weaken them as well and render them mutually dependent, science is above all a poorer and simpler thought. It is founded in immanence alone. Immanence must thus be deemed *radical*. Distinct from the philosoph-ical, since it no longer has the support of transcendence, it must be manifested (to) itself by remaining in itself or in its Identity. And this Identity has in fact no need to be separated from itself, to be sundered and alienated. There is not one, as ontology believes, but two heterogeneous modes of phenomenaliza-tion of the real. One is the circle of every philosophical Decision, in which the real distinguishes its phenomenalization from itself in order to circularly reidentify with it, and the other is the scientific, in which the real is from the start and remains identical to its phenomenality, which it is through and through. Unseparated from its phenomenality, the real has thus no need to pass through the transcendence of knowledge in order to be what it is. Unlike philosophy, science is not an alienation of the real, a real image of the real, and it does not continually derive from it. We will call this mode of "radical" phenomenalization, which suffices to define the real or its essence, the nonpo-sitional One (of) self or again the vision-in-One.

Yet, do we have an experience like that of the vision-in-One? A necessarily unconstitutable and unconstituted experience, describable in philosophical terms no doubt, but whose use would no longer be philosophical, i.e., categorial and essentially objectivating? Is there an experience such that, on its own and in the most immanent way, it postulates 1. the radical precession of its "object," of the real, on its description—thus the real as not constituted by its description but as absolute before its knowledge; 2. the corresponding possibility of a non-philosophical use of philosophy, of a noncategorial use of "categories" as a procedure of this description? This experience is the scientific source anterior to every philosophy and cannot be thought and determined by it in return. It can be neither "demonstrated" nor posited by ontological procedures, at the risk of falling back under the legislation of the philosophical Decision. Rather than the object of an acquisition by means of an operation of transcendence, it must be proof (of) itself. It must be given straightaway *as* nothing-but-itself, without surpassing itself, exceeding itself, hollowing itself out and transcending itself, without affecting itself, in a general way, with nothing, nihilation or alterity, etc. This is probably the *being-given* of the hypothesis and of the axiom, their non-philosophical or simply "posited" being.

How do we know that sciences are absolute (immanent, nonmetaphysical) knowings that always have, in a certain way, access to the real? From where do we extract this "idea" or "presupposition"? From what philosophy, perhaps, that we covertly presuppose while refusing to recognize it? To this philosophical question, we can only offer a scientific response. That is: rigorously immanent or transcendental. We must reply: sciences, through hypotheses and axioms, are *in-the-last-instance* an experience of the real-One and not only a theorico-technico-experimental assemblage. We know this from science itself and not from philosophy, which always refuses such a response and the right to make it. All that we know, we know from the thing itself and not from philosophical transcendence. We treat science itself as an immanent guiding thread of its theory because it is such, by its intimate constitution, that it knows in a scientific and not philosophical way that it is science rather than philosophy. It is *index sui (et philosophiae)* and furnishes on its own the means and above all else the immanent rules of its description.

It would be contradictory for us to claim, once more, to access the essence of science through approaches and procedures that stem from the philosophical Decision and that are impregnated with transcendence. This means that neither *reduction*, nor *analysis*, nor *meditation*, nor regressive *foundation*, nor the search for ultimate requisites, is still usable and legitimate. Nor the epistemological "reflection" on sciences' presumed "fact," which is merely an artifact of philosophical objectivation. Through what other technique can we then access the essence of sciences? Through no technique, for every technique always codetermines its ob-ject and coproduces it in the style of philosophy. No technique is necessary to situate oneself within the most general posture of science, through hypotheses and axioms, vis-à-vis the real. The real-One's essence rules out that it be what philosophy wants it to be: relative(-absolute). It may be that the Absolute has always appeared difficult to access, since philosophy was not constituted to think and experience it, but to denigrate, divide, and cast it into exteriority and transcendence, and thus to multiply the obstacles where none exist. Thought emancipates itself from its naive adherence to philocentrism and renounces the deforming prism of transcendence to the strict extent that it knows (itself) identical to this Identity-of-the-last-instance and knows this nonmetaphysical Absolute as the immanent being-given of the hypothesis and of the axiom.

Science is thus nondecisional science (of) self or (of) science. This knowledge is so radically immanent that it excludes philosophical reflexivity and its operations, all of which are tied to transcendence. In other words, science and the experience of science are the same thing or, more rigorously, are strictly identical in-the-last-instance alone. Science is the nonthetic unity (of) (self) (of) science and (of) the science (of) science.

Far from being formal and tautological, and far from expressing any Principle of Identity, these statements are nothing more than "transcendental theorems." They are theorems formulated by the science (of) science or *transcendental science*. They express the *real identity*, the identity (of) the real, and, finally, the identity of the last instance (of) the real and (of) knowledge (of) this real. Clearly the "foundation" that renders "empirical sciences" absolute in their own way is no longer by any means the logico-formal identity to which a reality would be attributed. This attribution would give place once more to a transcendental Illusion. It is a nothing-but-transcendental or real Identity that has never been *also* logical. Science—even formal logic, provided it is understood as an already authentic science—is the best destruction of the confusion of "general logic" with the real in the mixture of a "transcendental logic." This mixture is the heart of every philosophical decision. Science—even formal logic as science . . .—is an alogical knowing (of) the real, by its essence at least. It is not founded first on the Principle of Identity, which will have always served both as ground and as foil to the philosophical Decision. A science obviously cannot be reduced to this *identity*. In particular, it renders compatible an "objectivity" and an immanent reality of its *real objects*, those to which it relates these knowledges in-the-last-instance. Lastly, it orders or submits these latter, its *objects-of-knowledge*, to its real objects. But it is decisive for the moment—in order to distinguish science and philosophy—to identify precisely the purely immanent quasi relation of sciences (to) the real. This relation has never been a relation, since it is instead the immanence-of-the-last-instance alone (of) the real and (of) the knowledge (of) the real.

Of science we say: its *identity*. In effect, unlike philosophy, which always combines Identity with scission, transcendence, difference, and so forth, science discovers its cause, its ultimate reality in a *radical* Identity. By this we mean that it does not result from an analysis or a synthesis. It is without differentiation or identification; it is "in itself" and precedes all the phenomena of scission, of separation, of rupture, of interruption, of transcendence, etc. And so, even if it is transcendental, i.e., rigorously immanent to itself and received as such, it is in no way conflated with the superior Unity of the subject and of the object in the idealist way. This is why we ensure the real cause of science's objectivity in an Identity that is always already given before any opposition (for example: the identity of the subject and the object) *and never also concluded from them*. And this is why we aim to bracket, as philosophical still, the transcendental Idealism that continues to divide science's essence and thereby loses it.

Science is just a simple identity without identification. But it is at least, and at any rate, an identity. Its radicality means that it "begins" from itself alone and that it does not leave itself. Hence the term *nonthetic (of) self*, to signify that it need not alienate itself in order to posit itself. It is the identity of the real (and) of thought, but on the condition that we see in this identity their a priori root, *not a common root but a root of-the-last-instance*, nothing-but-anterior, rather than their simultaneously anterior and posterior synthesis.

The positive reason for all these phenomena, the reason that explains this (non-)relation (to) the real, can be summarized in this term: the vision-in-One. For us, it is an operative concept. Science's element is the One, not Being. Science is not a mode of the Western ontological project.

There are—they are their own criteria of reality and truth, transcendental criteria—*immanent givens in a nonthetic manner (of) self*, they reject objectivity's philosophical artifact. Thus science's essence is not given by philosophy, but experienced directly in itself and in the non-ontological form of the One as a radically unreflected, transcendental experience. In this experience we find the reality of a sphere of absolute immanent givens; we can immediately begin to describe them without having to proceed, as philosophy does, to the preliminary operations of dividing, constituting, reducing . . . the object. Our attitude, although transcendental, is rigorously naive, scientific, and not philosophical.

Since science is the sole mode of nondecisional and nonpositional thinking (of) self, philosophy, which *desires* this apositionality but *cannot* acquire it, denies science's autonomy. It is not only Kantian or neo-Kantian idealist epistemology that rejects the existence of immanent givens or of a nonthetic phenomenalization (of) self in order to oppose the *categorial objectivity* to them. It is the whole of philosophy that, as Decision or Transcendence, cannot truly access science's essence and produces in turn this "reactive" symptom called "epistemology." If the program of a rigorous science of philosophy passes at any rate through the "destruction" of epistemology, it is because the traditional unitary relation of prominence between science and philosophy is reversed and even more-than-reversed. For it cannot be a matter of a reversal of hierarchy and a passage to antiphilosophical "positivism." To sum up what is excluded here, we will say that science does not receive its essence from the philosophical Decision; that it possesses a positive and specific essence; that this essence does not allow itself to be thought as a mode of Being or an avatar of the ontological project or an exploitation of the properties of being; that, in general, its nonrelation (to) the real does not pass through the philosophical objectivity it absolutely precedes, and does not fall under the legislation of the ontological Difference.

THE IMMANENT ONE-MULTIPLE OF SCIENCE OR CHAOS

Whether it is understood in its substantialist sense, in its idealist-transcendental sense, or in the broadened sense of an ontotheology, ontology still lies at the foundation of all epistemological positions, even the most empiricist among them. It has never ceased to constitute the element in which the Occident has claimed to think the essence of science. Conceived in its broadest sense, Being has not ceased to be the ultimate reference, the last authority to which scientific knowing must be related and measured. It is this gesture that should perhaps be abandoned now in order to "test" another hypothesis, that of the One. What the Greco-Occidentals call the "One"—without ever distinguishing it absolutely from Being or thinking it for itself and in its essence—is obviously not the one that can serve to found science's postural realism. It may be that the forgetting of the One's essence and the forgetting of science's essence are the same thing, the same presupposition, which is necessary for the unbridled development of the philosophical disposition. But scientific thought, considered as such, elucidated in its essence, can be nothing other than the thought of the One or the vision-in-One. And just as we have posited the antiphilosophical rule: let science describe itself in its immanent phenomenality, so we have to prolong this emancipation and to suppose that the thought (of) the One is none other than the One-as-thought. We have to posit the following rule: let the One describe itself in its most immanent phenomenality. Here again, we suspect that the One thus experienced as nondecisional "vision" requires the exceptionless suspension of every philosophical Decision that is supposed or that supposes itself—as is always the case—co-constitutive of the real (of its sense, of its phenomenality, of its realization). This is why we will have to understand the One as One-Multiple and the One-Multiple as chaos. How should we proceed to these "rectifications"?

What we call the One is thus no longer Unity when it is mixed with the Dyad, interior and exterior to the Dyad, and with the Multiple. Philosophers' "Unity" is regularly connected to a Multiple, and this Multiple has another source and represents another origin of reality. So much so that these two principles, Unity and Dyad, are presumed to be given together for a thought of overhanging [surplomb] or survey, which is Unity itself, endowed with the power to transcend itself. This system requires a certain concept of the Multiple, which seems obvious and is no more elucidated than the One: the Multiple is necessarily *acquired* or *obtained* within the system via transcendence; it is a result, that of a scission, a nihilation, an opposition . . . of "contraries." It is at times the fruit of an interruption or a division, at others it is identical to a positive distance that serves as Unity. But it is always tied to an operation of transcendence; it is never an absolutely originary and immanent multiple that has its principle in itself. On the other hand, it is this nonoriginary that is posited as originary and constitutive of the real. As to Unity or Indivision, it too remains an abstract or formal principle. It has not been elucidated in its concrete or real essence, in its reception (of) itself; it is *supposed* nothing more than real. It is a simple function; it is required or requisitioned in a function of unification and synthesis of the manifold of "contraries" or the Multiple. The One and the Multiple thus abstracted from their reality form together a unique circular operation. For instance: a multiplicity that is neither One nor Multiple, in which the one is inseparable from the other, but where neither is elucidated in its real essence, where both are functionalized in an external and transcendent mode: philosophy.

We obviously should rid thought of this image, i.e., this surreptitious transcendence of the One and the Multiple. At least the image of their essence and no doubt, thereby, that of science. We will take care not to choose—this would constitute a new philosophical Decision—the One in the dogmatic-substantialist way, or rather the Multiple, as though it could be isolated from the One. This abstraction merely represses the other "contrary," which always finds the opportunity to return and to produce effects. In any case, in their philosophical as well as scientific use, they are inseparable. But while they are also separated in philosophy, once again two united contraries, a unified duality, they are strictly *identical* in science and do not form a new, more or less unified, duality. *Where there is no longer any transcendence or decision, there can still be some Multiple and even the most immanent, the most primitive multiple, that which will never have been the product of an operation of division or this division itself. And where there is this Multiple, there can also immediately be the One, i.e., immanence itself.*

In the laudable desire to found science on the most radical Multiple, we cannot believe that we can dismember the philosophical Decision's complex unity and simply *choose*, for example, as originary and as the real itself, the Multiple-without-Unity—which still presupposes transcendence as constitutive—rather than the Unity-without-Multiple. At any rate, the other contrary will come at the same time, but as unthought or repressed. It is a matter—this is an entirely different gesture, it is not a Decision—of no longer choosing the One or the Multiple *in their transcendent Unity*. It is a matter of considering their transcendence itself as already suspended; of allowing the radical Identity of the One and the Multiple to describe itself in its phenomenality, without founding itself in the transcendence of logos. For this transcendence is reduced or suspended by the description itself—and, despite appearances, this suspension, let's repeat, is not a choice or a transcendence. Here again, what we call the nonthetic One (of) self is no doubt still the One-Multiple—since to choose between "contraries" is no longer the problem or the method. But it is so in an Identity that this time excludes both Multiplicity and Unity in the form they have within transcendence, where they continue to be opposed as two distinct principles united in a synthesis or else in a unity. The One as real Identity is a *One-without-unity* (and) a *Multiple-without-multiplicity*. So long as we have not really rid the One of the ideal-transcendent Unity, and the Multiple of the Multiplicity that is also, that is still a universal-transcendent predicate, we remain shackled to the philosophical aporias of the Unity-of-contraries, to the unitary style of the "decision" and to abstractions of transcendence. The One in its essence, i.e., the One-Multiple when it is nothing-but-imma-nent, deprived of the predicates of Unity and of Multiplicity, is no longer a unity-of-contraries. It is what we also call the undivided, the individual [*individu*], the indivi-dual [*l'individu-a-l*], rather than indivi-duel [*indivi-duel*], or again *chaos* considered in its phenomenal and no longer philosoph-ical content. It is Identity that constitutes the basis of the reality of every thought and of every experience. The philosophical Decision clearly tries in its turn to divide this One-Multiple, to impose on it a division of intel-lectual labor, which should be specified in this case as a transcendent divi-sion of transcendental truth or of the real. And it does not separate the One and the Multiple without also separating the One from itself in the form of a Unity and the Multiple from itself in the form of a Multiplicity—both already incorporate transcendence and are thus partially constituted by it.

How can we describe the most real One, the nothing-but-One? How can we describe its immanence (to) itself and its immediate identity (to) the multiple, i.e., the One as nothing-but-individual? And in such a way that this individual is in the multiple state, is even the Multiple as the real root of transcendent multiplicities? In such a way that the Multiple can be the real basis or the cause of science?

Only philosophy can find here a paradox or a mystery. The great rule of the immanence of the description, the scientific and "realist" rather than philosophical rule, forces us to use philosophical or other objects, themes, operations, and languages (immanence has no "proper" language, which does not mean that it is inexpressible). But we have to use them precisely in terms of this immanence of the One, under its laws and in order to describe it without claiming to constitute it; in order to subject them as representations to this real that they can no longer claim to "realize." This scientific description of science itself is indeed the description of a specific mode of phenomenalization (of) the real. But this description is subordinated to the real or presupposes it as an absolute precession, without also claiming at the same time to codetermine and thereby to transform it. Manifesting the essence of the One scientifically, and by the same token the essence of science, we reveal it in a way that is immanent (to) itself without transforming it in this operation. We simply transform the content of representations or of "knowledges" produced of this essence.

Other possible descriptions of the One, more complete perhaps, were carried out elsewhere (*A Biography of Ordinary Man, Philosophy and Non-philosophy*). We will not repeat them here.

SCIENCE'S "POSTURAL" REALISM OR THE INVISIBLE MANIFEST

The technico-experimental apparatus is a material and a means ordered to the essence of science. It is not this essence itself. Science's essence resides only in positivity, the quasi-"ontological" consistency of naive and "decision-less" realism and certainty, which affect the theory itself and its particular means. To be sure, it is not a matter of local "objects" and "representations" produced by science, but of what every scientific posture immanently postulates about the real to which it is related as such—of the scientific "intention" and its *transcendental claim*, if you want. So this conception of science does not rest in any way on the Husserlian principles it neverthe-less evokes, and even less on "occidental metaphysics." It rests on a renewed experience of what is given in an emergent and irreducible (if not certain) way in the hypotheses and the axioms: *on an experience of the lived real as radical in its type of transcendental immanence.* Furthermore, philosophy's traditional—and necessary—skepticism about the existence of absolute immanent givens (even Husserl limits their reality in his transcendental Ego, which he partially obtained through philosophical operations) is not a serious theoretical objection for us. It is a simple "resistance," founded in the philosophical Decision as *transcendence,* and it uses theoretical means; science can analyze it as such.

Science is realist in the sense that it knows itself straightaway as science (of) the real. This fact does not rule out, but rather requires, a rectification and a recasting of representations. That it is immediately *identical* (to) the real and to its fundamental immanence, without having to pass through a process of identification with a transcendent real, should not be conflated with a mysticism that excludes representation. Representation is quite simply not the foundation of sciences. Sciences are not sciences—rather than a mixture of already existing sciences and of philosophy—unless they assume this realist posture of the last instance: above all, not the eventual reality-effect which this or that particular representation can induce. From this standpoint, science implies the destruction of the primitive transcendent realism of perception. It is a matter of the immanent or transcendental reference (to) the real of-the-last-instance, a reference of the theory that appears with the index of exteriority and transcendence. Science establishes the distinction between a transcendent realism it destroys; a transcendental yet objective, therefore ideal and semitranscendental realism it partially destroys, only conserving transcendence in a simple and nonthetic form, and a nothing-but-transcendental or "postural" realism in which the real is immanent, even as "object." If science has an "ontology," it is neither empirical-substantialist nor idealist-categorial nor existential and projective. It is strictly the ontology of immanent phenomenal givens, which are what is known *in-the-last-instance* by knowledges.

This postural or immanent realism clarifies what remains philosophy's stumbling block: the Identity (of) the real, the vision-in-One, is deprived of transcendence and of logos (of reflection, light, position, horizon, project, etc.); it is nevertheless a thought or an experience, a nonthetic knowing (of) self. Science's lack of consciousness or reflexivity, its obscurity, and its blind nature do not signify the absence of thought as philosophy believes. This is one of the most formidable Greco-Occidental misunderstandings. There is indeed an opacity of scientific thought, and this thought does not think like philosophy. But it "thinks" in an original and positive manner. Unable to understand the sense and origin of science, which is the nonthetic Identity (of) thought and (of) the real, philosophy falsifies all those traits by interpreting them as a lack or defect of reflection, a degradation or a deficiency of the light of logos, of reason itself . . . This confusion of the scientific order and of the philosophical order leads to the philosophical denigration of science. Considered as a mode of the ontological project or else a mode of self-consciousness, science becomes a necessarily dethroned and deficient mode. It is condemned to discover in philosophy the sense, the value, and the truth it naturally lacks . . .

In reality, it is quite possible to "reconcile" in the vision-in-One not only knowledge but also thought and even the transcendental with science's unreflection. It is enough to conceive this unreflection, not as a lack of logos, but as the positive structure of the vision-in-One, which is indeed a thought or a transcendental experience, even if it is without distance to itself or exteriority. The whole strategy consists in inverting the sense and the origin of this apparent scientific nonthought. Instead of imagining it the way philosophers do, as a supplement or an excess of objectivity, as an objective yet brute reality projected beyond its condition (the dimension of objectivation), we will bring this scientific realism under objectivation and objectivity themselves. We will seek science's apparent non-sense in its original essence of the vision-in-One. This essence excludes sense, but through excess of reality rather than through excess of objectivity; it excludes transcendence not through excess of transcendence, but simply through immanence.

Philosophy seeks and posits science always too far: at the end of its "reflection," of its "project" of objectivity, of its "dialectic"—in general, at the end of the transcendence that founds all these techniques. It is precisely this transcendence that science excludes at least from the relation (the nonrelation) it "entertains" with the real of-the-last-instance. Hence science's naïveté, unreflection, realism, and "blindness," which are so intolerable to philosophical objectivation that it has not ceased to deny, reduce, and falsify them. This is what is called epistemology; it is even the *epistemo-logos* of every epistemology.

Science's opacity, blindness, and muteness, as well its realism: these phenomena are not those of the supposedly brute and meaningless object. Rather, they are the phenomena of thought itself, of the transcendental experience when it is, as is the case here, "unreflected." Only absolute immanent givens explain the paradox of these phenomena, which, on the mirror of philosophical representation, are reinterpreted as defects or insufficiencies (genealogy of philosophical judgment about science). By right, these *absolutely immanent givens* cannot be grasped by philosophy, which is founded on transcendence. But they explain the fundamental scientific realism, the "dual"—i.e., without synthesis—distinction, science's nonconfusion of the real and of objectivation.

Thought in its real essence is thus invisible within the horizon of objectivity, on account of its structure of radical immanence, of the immanence that remains in itself. The vision-in-One is invisible in the sphere of the Greek presuppositions of Being or of logos and the Judaizing presupposition of the Other. It is manifested, we might say, only through this invisibility that must be understood as a thoroughly positive essence rather than an oblivion, a withdrawal, a "self-concealment" of the phenomeno-*logical* type in order not to cast it into a transcendence. How can we grasp this indivisibility in a positive way?

The true manifest is invisible in its very essence or is the essence of this invisibility. The essence of the real or of the radically immanent phenomenon is the invisible that has become positive and is finally received as such in its own mode, which is invisibility. Invisibility ceases thus to be the negation, lack, or privation of an ontological or phenomenological visible. The Invisible is no longer an attribute or the predicate of an ontic subject interior to Being's transcendent visibility, nor is it the predicate or property of Being, elevated to the state of Being's essence. The Invisible is, from the start, "essence," without having to be first the attribute of a subject. The real One is nonpositional (of) self: it is not alienated in an attribute, any more than it receives its sense from an attribute or is determined by a universal. The task is thus to tear the Invisible away from its state as an attribute in order to describe it as a "subject" without attributes. It will perfectly visible as such and its invisibility-of-structure will not be undermined, limited, or even partially negated by its visibility, only if it ceases to be experienced in the transcendent frame of perception, then of ontology, in the general context of the unity-of-contraries. Because the Invisible is nonthetic "cause" (of) self or essence, it has no contrary. Even its specific visibility, its own phenomenalization is not its contrary or what limits it, but forms a type of manifestation that suits the Invisible and safeguards it as such.

As to scientific discourse itself, to produced knowledges, it is not, it has never been, an illumination that transforms the real. It is a simple *reflection-without-mirror*, a unilateral manifestation. It remains invisible by its essence of-the-last-instance, but renders the real visible in this mode of nonthetic reflection, without making this visibility penetrate into the essence of manifestation and without transforming it in turn. Scientific representation is the manifest. But since this manifest has the Invisible (of) the real as its essence, it does not render this Invisible ontologically or epistemologically visible: *the Manifest is invisible—it is the real; the Invisible is manifest—it is the science of the real.*

Philosophy is caught in this trap. It fails to grasp this phenomenality, more interior (to) self than every philosophical subject, which is already partially tied to a transcendent attribute. And it interprets the phenomenality by projecting its own model of the object on this real. It is then condemned to imagine the real as an absurd beyond of its own object. If so-called empirical sciences do not seem to correspond to this picture, it is because we continue to examine them through the philosophical prism of objectivation, as a degraded or dethroned form of objectivation: a hyperobjectivity. An immanent, purely phenomenal description, disabused of philosophical prejudices, shows, on the contrary, that science derives its specificity and its autonomy from the fact that its naïveté, far from being a deficiency in philosophy, in reflexivity, and in self-consciousness, is its essential and positive structure. It gives science its consistency. Just as the One's essence was "forgotten" by philosophy in the name of the metaphysical "One," which is always transcendent in some part of itself, just as it was requisitioned and put in the service of the guardianship of Being, so science's essence was "forgotten" or denigrated by philosophy. Philosophy conflated science with its own operations, with the project or the objectivation; it mistakenly assigned to it the unique task of knowing the object, accused it of lacking this knowledge or of lacking the problem of the ob-ject's origin, and ended up relegating it outside the experience of the authentic real under the now inevitable pretext of naïveté, technicism, manipulative and blind thought, and so forth. Science's naïveté is real, but it is essential and positive: it is that of the transcendental nonreflexive or nonthetic experience (of) self. Science "thinks"; it *therefore* does not think in the philosophical mode of transcendence or of position. It has no ob-ject in this sense; it is vision-in-One.

HOW TO THINK IDENTITY QUA IDENTITY: THE HYPOTHESIS OR THE AXIOM OF THE LAST-INSTANCE

"To think Identity qua Identity" cannot have the sense of an ontological injunction, or of a call to: "take care of being as a whole," nor the quasi-ethical sense of a commandment or of an imperative to "be the keeper of the Other man as of your brother." It is an *axiom*, a simple position of immanent thought, for a thought that would equally be a science. It is the description of a posture of thought that is adopted right now without preconditions or philosophical presuppositions.

A scientific posture confines itself to starting from the "object" to be explained, treated as a guiding thread, or as a cause of the new representation, a cause that is not exhausted in this representation. But it undertakes to modify representation in its existing forms in terms of this new object. The scientific posture, as we have described it, programs the abandonment of philosophical operations and decisions as constitutive. But it does not program the *rejection* of philosophy, which subsists in the state of "objective givens" necessary for the construction of the new theoretical field. This task is obviously complex and presupposes several kinds of operations that will be described later on, but its sense is clear.

Whereas philosophy takes as its transcendental guide of description, in the worst cases the constituted object or else objectivation itself, and in the best their difference—Differe(a)nce—i.e., in all cases, some *transcendence* (exteriority, nothingness, scission, etc.), science takes as its guide the "real object" to be known. This object is never given in the form of object-objectivation, in the form of transcendence or of representation in the state of autoposition. *And Identity is given in this way least of all: for Identity is not any object whatever—as we might expect—but the "prototype" of every real object.* Since it is no longer possible to *objectivate* Identity and its causality in any mode whatsoever, we have to *let* it think itself; to stop hanging over it philosophically and putting ourselves in its place; to accompany instead, at its own level, Identity's experience of thought and to confine ourselves to describing it, to manifesting it *as such*, without claiming to constitute or transform it through this operation, which must be a pure "reflection-without-mirror."

Now, how does a transcendental science—transcendental both because it is real science (of) the real and because it knows (itself) to be such—think while producing knowledges? The formula "a science of Identity qua Identity" is equivalent to the formula "the One qua One." Both can clearly receive an ontological or philosophical sense, like the other statements about Identity or the One that accompany them in this context. But can they receive a truly scientific sense, which would no longer be the sense of the Aristotelian "science of the One" that had to recognize its failure as science?

The hallmark of a philosophical statement is that its sense as a statement and its nature as an enunciation continually and circularly communicate—save for a few nuances that are impertinent here—and that a statement like "Identity qua Identity" is presumed to be endowed with a will or a power of autointerpretation and autoposition. To treat it, on the other hand, as a scientific statement, two conditions must be fulfilled: 1. This statement and the others must stop interpreting themselves; they must stop transmitting a knowing, and the sole possible knowing, about their object—that of which they speak, Identity. They are now simple *data*, indications for the production of a new knowing—which they do not already contain in any capacity—about the One. These statements can thus be explained by something other than themselves, other than their "spontaneous" philosophical economy (syntax and sense). Here the rule is that the real explains its representation, not that this representation explains itself by becoming explicit. The latent "hermeneutism" of every philosophy is eliminated: the couple statement-enunciation is no longer enough to produce some knowing; it requires the radical and unilateral distinction between the real (the object to be explained) and its representation, with the real functioning as the cause—but of-the-last-instance—of representation. The real cannot be deduced from representation by any theoretical procedure; it is representation that is deducible from the real by means of a certain labor carried out only on this representation. The real-cause is not circularly exhausted in its representation. It is not determined by this representation, but rather determines it in the form of a new object-of-knowledge. 2. But the most significant condition is the following: we can only treat Identity as the real object of a science by positing it as a theoretical "object," radically new vis-à-vis the state of the existing or present theoretical fields—in this case, the field of philosophy, which seeks to be but is not or is "not yet" a theory. If the cause that explains scientific representation is the real, insofar as it unilaterally or unequally determines representation (the very sense of "last-instance"), and if the real is so original in relation to representation that it is not even its "Other" but its "cause," the cause that precedes it "unequally," then objects necessarily emerge as absolutely new in the given theoretical field. And science is nothing other than the labor that produces at last an adequate representation of these objects. We must thus treat Identity as a theoretical object of the type "hypothesis" or "axiom" (we do not distinguish the two here), philosophically unheard-of, ontologically unintelligible, and capable of entirely displacing this spontaneous theory that is philosophy. That the statement "Identity qua Identity" was first formulated in ontology does not affect this consequence and proves all the more that ontology as a whole must be treated as a "symptom" for science (we will employ other concepts: "indication," "material," "symbolic support"). Ontology is only apparently first in the order of knowing, but in reality it is present only as a material. That we treat Identity as an emergent object from which thought sets out, and not as a simple presupposition to be made explicit or to be understood and critiqued, has also nothing exorbitant about it. For philosophy gives itself in the manner of a first Given (but not exclusively, this is its illusion), with Difference for example as its experience.

How can we now think this Identity qua Identity from the standpoint of its materiality? For this philosophically impossible object, we have merely ready-made spontaneous representations, already available in an existing corpus, but in fact inadequate. We can already eliminate some typical and traditional representations of philosophy: the form-object, objectivation, the intentional aim, every form of transcendence that pretends to aim for Identity and that could only divide it anew and negate it. Rather than an *ob-ject* in the philosophical sense, it is a cause, and the true "objects" of science are *causes* rather than *objects*. But here again philosophy offers its spontaneous representations of the cause, which must be discarded. It separates the cause (its Identity) from what it can do. It distinguishes four forms of the cause (form, agent, end, matter) and only recognizes to them a unity by analogy or by other procedures that claim to reconstitute the identity of the cause from the outside: philosophy loses the efficacy of the cause *as cause* or again the cause of the "last-instance" and disperses it in the transcendence of Being. The forms are clearly modes of Being, and it is thus that the essence of causality (Identity) is lost for the benefit of its convertibility with the transcendent, illusory "causality" of *Being*. The minimal, irreducible efficacy of the cause as cause does not pass through the mediation of form, end, agent, or matter. It is rooted in the immanence that constitutes Identity—the cause that is not "immanent," but is the very causality of real immanence. But how can we think in a rigorous theoretical mode the whole of these determinations, for which philosophy has no instruments and on the very denial of which it is constructed?

"To think Identity" means: to think it *qua Identity*. And this implies: to think *according* to the *already given* Identity, in terms of this Identity as the very cause of thought rather than as the object that is aimed for. This fact has already been established. But having eliminated the *form-object* from the cause in its nondecisional, nonontological, or nontranscendent form (of) self, we have to equally eliminate the *form-subject* from science. Identity as cause (of) self of-the-last-instance is not a transcendental subject any more than it is an object. For the philosophical "subject" is not only the object's *correlate*; it is a mode of this latter, equally divided or shackled in its essence to the division, and its essence is thus abstract or dependent. Science is a process-"in-cause," nonintentional cause (of) self. And if this process is *without-subject* or without will (for it the deconstruction of metaphysical will is an obsolete question), it is also—one forgets too often as a result of materialism—*without-object*.

Thus *to think Identity means first of all: to think (the World) according to Identity. But a thought that thinks "according to Identity" thinks Identity itself only to the extent that, not treating it as object or as subject, as mode of transcendence in general or of Difference, it thinks in the "succession" or "posterity," in the inequality of Identity as "last-instance."* Since Identity is phenomenal or manifest through and through, it has neither need nor desire for an operation to support its manifestation. By definition it will have never been a metaphysical *will*, a subject, an object, or their Difference, and has no need to be deconstructed. In effect, the *qua* no longer indicates here the donation . . . by means of Being, but the sole Given that has the power, being given, to transform the operations that would claim to give it; to strip from them this claim in particular. What has *to be thought* is not some more or less forgotten already-thought or some unthought that remains to be thought; it is some *given* that has no other character, in any case not the "idealist" character of being some already-thought or some yet-unthought. The most rigorous thought does not think thought (old or implicit and concealed or self-withdrawing); it thinks some *given*. Here the term *given* does not mean "primary" and "foreign" to thought or its Other, but—this is altogether different—what precedes it in principle and forever as its cause. And the cause that unilaterally derives it is not the Other of thought, but transforms thought into its Other.

We have managed to treat the One as a first Given, an emergent theoretical object, only because it is clearly constituted of immanence alone; it is not blended with any transcendence or representation. And when it is a matter of Identity, "immanence" is not a simple predicate; it is the essence or the real itself, which knows (itself) as real without passing through the mediation of representation. Real Identity, what we call the nondecisional cause (of) self, is no longer that which is *convertible* (with Being, Difference, the Other, the analytical relation). It is no longer a mode or even an ingredient of Difference. It is that which is "qua Identity." We will call it "real Identity" in opposition to its logical form (the "*Principle* of Identity") or even its logico-real form (Self = Self). The "qua" does not indicate mediation by division, thus by Being and soon by Difference—the Difference that Being always is and the Difference that will soon be the *essence* of Being. Instead, it indicates the being-*identical* or the *being-identity (of) Identity*, its immanence, which is anterior to every scission and forever inaccessible to such a division.

To think Identity qua Identity now also means: to treat the traditional philosophical statements about Identity (the whole ontology and metaphysics of the One, for example) as objective givens or phenomena of this thought *and to relate them to Identity as cause-of-the-last-instance*. From this redistribution of Identity between the determination-in-the-last-instance and the determinable emerges a new experience of thought, a scientific reform of the understanding. To transform the indicative statements supplied by philosophy in terms of this cause, which is not in its turn determined by this material: these specifications eliminate every risk of objectivating Identity and thereby of "subjectivating" it, every temptation to turn it into a superior object or an ultimate subject. But they also preclude its other convertibility, its other equally philosophical confusion: with the Other, after its confusion with Being; with the Trace, Withdrawal, Difference, after its confusion with the Object. *Qua,* in ontology, designates an attribution, a sense of . . . and for . . . , as a predicate given to Identity through a necessary operation of language. In a science, by contrast, it directly designates the essence itself, that which does not depend on an operation of language; the cause that instead conditions language. Science destroys the predication's traditional ontological scope and reduces it to the secondary function of materials.

Once Identity is posited as *a scientific discovery that globally exceeds the philosophical horizon*, the only problem is to recast this false theory that is philosophy. Our work only uses the material provided by philosophy, but its principles stem from science alone. In philosophy and for several reasons, Identity alone, which is not convertible with Difference, appears unthinkable. But science does not preoccupy itself with this prohibition cast on Identity, and, just as it treated Identity as a theoretical discovery, so it turns the adequate thought of the One into a simple problem it has the means to resolve. For science, it is a matter of discovering, if not "inventing," the forms of the specific thought of emergent Identity; and of doing so by rendering philosophy and its authority immediately contingent. Science does not try to deduce these forms from philosophy's operations as though philosophy constituted the unsurpassable horizon of every thought. And so it ceases to do ontology, to pose aporetic *questions* to Identity, to suppose that the philosophical use of language and of representation is the only possible one and to *object* to the Idea of such a science.

Furthermore, Identity as such sets us on the path of a "third type" of experience of thought, which is neither that of Being nor of the Other. A thought has to be invented, or rather discovered, that does justice to this philosophically incomprehensible exigency: that Identity is thinkable *in-the-last-instance* (not in the manner of a circle, of course) *on its own, with the simple help or the occasion of philosophy rather than thinkable by philosophy*. Provided it is the very Given that precedes every operation of thought or of constitution, Identity has the force—it is this force—of defining a new form of thought, which is unilaterally ordered to it and cannot be discovered within the horizon of philosophy. Philosophy does not modify its operations or its posture in terms of the real, but claims to impose them on it, and, for that reason, it begins by alienating the real, raising it to the surface so as to better objectivate and cut it willingly or forcibly. This languid voluntarism is philosophy itself, which insists or prefers to "endure" the real rather than to let itself be modified or transformed by the real. A science, on the other hand, ceases to languish and only transforms the object, which it distinguishes from the real precisely because it paradoxically begins by letting the real be-*given*.

The Theory of Identities-of-the-last-instance is thus not a transcendent or arbitrary theory, a "doctrine," a new "philosophical position." It is not the false theory and the true ideology of ready-made Identities, conflated with phenomena or else abstracted from them, elevated above "experience." Its "objects" in the most general sense of this term belong from the outset to two heterogeneous types (this averts every confusion): on one hand, philosophical statements about Identity, which now serve it as *data* or *phenomena*; on the other, and especially, Identity as cause of the regulated *theoretical transformation* of these statements. In both cases Identities intervene insofar as they are "posited" or required by a science. We will take care not to imagine the Identity-cause in the manner of "natural," "geometric," or else metaphysical singularities. These abstract identities of phenomena are, in reality, *divisible* like the phenomena themselves and can at most serve as *data* in this science. If thoroughly undivided Identities are already given (and, if they have to play the role of the cause-of-the-last-instance, they can only be given unconditionally), then they are not given in nature, in dynamic geometry, or in philosophy, *but in science itself and as such rather than in its regional "objects" or its philosophical Idea.* They are indeed what we can call "theoretical objects," discovered or given as such; but they cannot in fact be discovered on the plane of constituted knowledge, which has become just as transcendent as a "nature" or a philosophy. Instead, they are given as what forces the theory (of these objects) to modify itself.

THE THOUGHT OF IDENTITY QUA IDENTITY: A SCIENCE RATHER THAN A PHILOSOPHY

Although these are indications to be elaborated, several traits of this thought at least suggest a science, which can and must be named in this way:

• The only causality required here is the causality of the real as strictly immanent: we can call "science" the knowledge grounded in reality. This causality of radical immanence, on the other hand, implies the immanence of theory's real criteria and of theoretical criteria themselves. We can call "science" a knowledge grounded in rigor and in autonomy.

• This thought is the full exercise of causality, i.e., of a nondivided causality. The determination-of-the-last-instance is the cause qua cause, whose efficacy is not divided between it and its cause and does not return to it through its cause. Thought as theory, for its part, is the effect, as effect (which knows itself to be an effect and nothing more), of this cause and does not claim to codetermine the cause in return but lets-it-be as cause. We can call "science" a knowledge that is unilaterally ordered to the real.

• The thought of Identity as the sole (real) cause is the simplest of all and the most minimal. The most rigorous, the one that reflects rather than expresses the real's structure with the least possible mediation and that emphasizes this structure as the cause of theoretical representations. The poorest because it rejects—although it exercises itself as a transcendence or a dimension of the theoretical—every foundation in a metaphysical or religious transcendence: in an autoposition.

• It is a reflection-thought (of the real); but an unreflected reflection, without the mediating structure of the mirror or of the third term presupposed by every philosophy. It is ordered without remainder to the real, but in-the-last-instance alone, as the absence of the mirror-mediation requires it. And it submits itself to the real of which it is the reflection by ceaselessly producing new knowledges that do not determine this real.

• Its materials consist of *representations with theoretical claims* (philosophies) and of already elaborated experimental *data*, which are reified knowledges . . . rather than alleged properties of things. It is not empiricist but "materialist" (in the sole sense that it uses *materials*).

• It is not a science of the "singular" (there is only a philosophy of this "singular," which has as its object the "difference" between the singular and the universal). It is rather a science of Identity, by means of its transcendent (philosophical) *phenomena* (for example: "singularities"), which are symptoms or indications for its universal theoretical representation. But the problem of a *singularity of-the-last-instance* will be posed apropos this theoretical universality itself.

• Philosophy *aims for* Identity, but in reality it thinks Identity in a different way than it aims for it—precisely because it merely aims for or intends it. It thinks Being or the Universal rather than Identity and thus produces symptoms. Science thinks Identity just as it "aims" for it. More explicitly: because it does not aim for it, but thinks (it) starting from itself, i.e., rectifies the "symptomatic" indications of Identity that philosophy supplies. Philosophy exploits and displaces from within the symptom that it *is* by its very existence. In contrast, science unravels the symptom: not in itself or by claiming to destroy it, but *in* knowledge (of the philosophical symptom) and on its behalf. We can call "science" the knowledge of the symptom that is not itself symptomatic, even if it becomes so once again for another knowledge.

• If there is a *problem* to be resolved, it does not affect the very phenomenon of the Identity-cause or the Given, but what is deduced from it as well as from this Given. Science's cause-of-the-last-instance is, in effect, announced "within" theory or is manifested within the sphere of knowledge in the form of the existence of a *new theoretical object*. This object requires existing thought to stop claiming to be valid for it and to transform itself into a simple material or *data* for a specifically theoretical recasting of these knowledges. We can call "science" a knowledge that presents itself as the resolution of a problem or in terms of a new given theoretical object. That science has a cause in the real, a cause distinct from itself in the mode of the "last instance," is enough to order current knowledge to the theoretical form, i.e., to the requirement of a new and more universal knowledge, which begins by invalidating or limiting the old one to the state of *datum* (of "material") and proceeds through the inauguration of a new discursivity. For example, at the most elaborated level, a non-philosophical or nonphenomenological . . . discursivity.

SCIENCE: A REALISM OF-THE-LAST-INSTANCE; THE SCIENTIFIC DESTRUCTION OF BACK-WORLDS AND OF PHILOSOPHICAL TELEOLOGIES

The revelation of Identity qua Identity takes place in the form of a new and particular science. Inversely, Identity is what allows us to rectify our idea of science "in general" and to wrench it away from the philosophical horizon. We will call "science" the manner of thinking that relates phenomena to their Identity as their cause of-the-last-instance and does so by means of the theoretical representation (of) this cause. A science is the theoretical knowledge not of phenomena, but of their cause (the Identity of the real) by means or the "occasion" of these phenomena. "Knowledge-of-phenomena" is an ambiguous, amphibological expression that has a philosophical inspiration; it suggests that science is *fundamentally* an activity of objectivation and a knowledge of objects in the philosophical sense of the term. In reality, the real alone is known, the Identity (of) phenomena rather than phenomena themselves. We will not conflate the theoretical objectivation of phenomena with their knowledge. The former is the means of the latter; knowledge is knowledge (of) the real by means of this objectivation. Knowledge on its own does not have any real meaning, but only a meaning as representation. The actualization of Identity qua Identity compels us not to conflate the real and its knowledge amphibologically.

How should we now understand the formula: "Identity of phenomena," which says the real object of every science? In philosophy, it signifies a continuity, a blending of given phenomena X, Y, Z and of their reality or Identity. This blending has no meaning for science, which gives another sense to the formula and teaches us to distinguish the "objective data" of a theory and their Identity. It dispels their spontaneous confusion and assembles them instead on the mode and by means of theory in new, philosophically unintelligible relations. Instead of their *identification* in an amphibology fated to autoposit and autointerpret itself, science arranges them in a process of knowledge. This process presupposes operations that exclude every autointerpretation and that afford this relation its reality. Philosophy's transcendental illusion consists in believing that Identity can be discovered at the level of objective givens, in their extension or at their horizon, in the mode of transcendence in which Identity is divided. But such a phenomena or objective data of a scientific field is only said of Identity in the mode of means-of-knowledge and insofar as it fulfills indicative functions for a certain region of data (material, index, support), but not functions that constitute a real object. And Identity is only said of the "objective" phenomena under the reason of the "last-instance," of the cause (of) self that determines without amphibological continuity. There is indeed a relation between the two, but it is a relation of knowledge, marked by contingency from the perspective of the reality of Identity or of its internal constitution. The "objective givens" (old theories, laws and models of the domain, technico-experimental labor) are necessary only to the scientific representation of Identity, not to Identity itself. Still, this necessity is of a special type, given the apriority and specificity of the scientific dimension, which cannot be reduced, as we will show further on, to its local theorico-experimental components.

Common sense and the philosophy that takes over from it spontaneously presuppose a continuity of nature and sense from the objective givens to the Identity (of) the real. This is how we speak, without reflecting on it anymore, of the identity *of* a people, *of* a race, *of* an object. Scientific practice sufficiently suggests that any phenomenon whatsoever is indefinitely divisible and multipliable in its properties and that if there is one Identity of these properties, it will not be on the same plane of reality as they are, but will have to belong to another order. Obviously the paradox that an analysis of the philosophical decision (one that does not come to a stop prematurely) identifies is that Identity becomes a metaphysical *back-world* [*arrière-monde*] of phenomena the moment it is placed within the continuity of "empirical" phenomena, in a relation of essence (in the metaphysical sense) or of difference to them. But the *difference* between undivided Identity and divisible phenomena is only the last possible form of back-worlds, whose concept is inseparable from philosophy's.

The complementary paradox—always for philosophy—is that the scientific "separation" of phenomena and of the Identity (of) the real, their unilateral or unequal distinction, is precisely the destruction of every possible back-world. Identity as such appears to be a metaphysical "interiority." But if it is *absolutely* primary and not divided from the beginning, if it is nonintentional cause (of) self, if science installs itself within Identity and remains there even when it assembles the elements of its theoretical representation, it is because there is no longer, there has never been any back-world: so wills the "last-instance" . . .

Identity qua Identity is the destruction of metaphysical worlds, of the "World" that is always a "back-world." The only way to vanquish metaphysics (and its scientistic and positivist species) is to situate oneself in the posture of radical immanence of the One, which is neither a Beyond nor even an Other of the World, but the cause-of-the-last-instance that enjoys [*jouit (de)*] its precession on the World. Instead, it is phenomena and, in general, "objective givens" that are "beyond" the Identity (of) the real. This is more than a reversal of the metaphysical hypothesis. Science confines itself to thinking according to the real order. Not, like philosophy, according to the order of *supposedly* real phenomena, but to the order that goes from the real *to* phenomena—causality itself. The indefinite scientific divisibility of properties shows that there are no simple natures or essences at their level, in *correlation* with them, whatever the ("differentiated") mode of this correlation may be. Nothing can stop the scientific analysis of phenomena, their dissolution, but the analysis never reaches anything real. It is a procedure necessary to theory, insofar as theory is a nonobjectivating knowledge *of* the real. Science presents and describes Identities: the Identity of Identity as such, the Identity of other orders (Theory, Experience, or again Being, being, etc.).

The concentration of the real outside phenomena, within Identity alone, opens a new type of space for thought: *theory*, which, unlike philosophy, is representation as determined in-the-last-instance by Identity as cause (of) self. Theory is the sole thought *adequate* to Identity precisely because it no longer objectivates the real à la metaphysics. But why this liberation of thought? From the viewpoint of reality, the concentration of the real in a cause-of-the-last-instance renders contingent, first, the "objective givens" or the "phenomena" and, then, the theory that uses them. So much so that the theoretical space is no longer itself *intrinsically limited in its very reality* (which is a priori and dependent only on its cause-One) by these *data* of experience, which are necessary to it in another capacity, that of "simple" indication and support. Hence an absolute, infinite de jure opening of theoretical labor, the production of knowledges that encounter no ontological limit, no philosophical finality. The finitude of Identity, radical immanence, is clearly the reason for this free opening of knowledge as thought. Correlatively, since "phenomena" cease in their turn to be predetermined by Being or by some ontological finality and horizon, and since "experience" is delivered to its own *identity*-of-order, science is extended to any phenomenon that can now become the "object" of a science. Or, rather, that can *give place and indication to a new science*. Science's open multiplicity has no other reason than this expansion, for which the Identity (of) the real is the cause and militates against every transcendent philosophical economy of "objective givens" and of the field of research.

The ensemble of these effects, which are produced on thought and its relation to the World, signals a *scientific derealization of the World and of philosophy*. Science now makes use of these latter in the nonontological capacity of simple index, material, and support. A nonmetaphysical, non-Cartesian derealization. They are not "simple" phenomena abandoned to science; it is, on the contrary, science itself as authentic thought, endowed with a transcendental yet no longer metaphysical power, that derealizes the object and abandons it to philosophy itself. More than ever, in a sense, the "method" (in this case, the scientific posture itself) constitutes—a nonidealist, non-Kantian constitution this time—its "objects" as simple givens in terms of indication and support and does not allow the World's demands to be imposed.

SCIENCE'S REAL OBJECT OR ITS A PRIORI STRUCTURES

Now that this first description, the main one, has been completed, it remains for us to describe the other phenomenal givens that constitute the *reality* of a science. First, those of the *real object* in the narrow sense, to which a science relates its knowledge. Nonthetic Identity founds a quadruple a priori postulation of reality, of exteriority, of stability, and of unity that together constitute the *real object*, the a priori condition of the object of knowledge. It founds here the noncircular relation, free of reciprocal determination, that makes it so that a knowledge is subject to the real and only claims to "reflect" or describe it through the very operation of theorico-experimental production of its representations. It finally founds the *object of knowledge* as the articulation of empirical procedures (in the broad sense: the whole theorico-technico-experimental apparatus) qua phenomenal objective givens and aprioric procedures of the real object (the insertion of this apparatus and its labor into the real object, in view of knowledge). It is then a matter of understanding how the "epistemological" givens and the special realism that science postulates are articulated. The distinction between two objects does not, in effect, encompass the distinction between experience and the concept, the concrete and the abstract, experimentation and the theoretical—or any of their "dialectizations" or "couplings." The real object already contains theorico-technico-experimental ingredients (but in the state of overdetermination of a priori structures). So much so that the two objects *contain the same representations, but with an altogether different status.* Their distinction is not epistemological (or transcendent, from our point of view), but uniquely of-the-last-instance, i.e., *transcendental or immanent* and founded in the very essence of science.

To begin with, science has to specify the knowledge of the real-One in the form of an a priori quadruple structure called the "real object." If the "real object" expresses a transcendental claim, for example a claim of reality and of stability, which is that of science, it does so in a mode that equally shows the feeling of exteriority necessary to science. Although founded as absolute (albeit finite) power on a pure immanence that does not contain the least parcel of transcendence, science does not exclude transcendence absolutely. Quite the contrary, it requires that the transcendence of its objects and of its representations participate in this absolute essence or this wholly immanent phenomenality. Because it blends the essences, philosophy believes that this requirement is contradictory and proposes itself as the solution to this "contradiction," which it resolves by mixing or coupling contraries. In reality, it is entirely possible to describe, among the phenomenal givens that constitute a science as such, a special experience (of) transcendence. A nonautopositional experience, unknown to the philosophical order, because it is a form of objectivity that flows in-the-last-instance, in an irreversible order, from the One; it is not viciously transferred or somehow copied from the experience of objects or from their objectivation the way it is in philosophy, which always "fabricates" some transcendence with some transcendent. There is—given immanently with the One for a rigorous description—an experience of reality, of stability, of unity, and of exteriority; but they are also nonthetic (of) self by their essence, which is the One and does not produce their autoposition: this is not a de-cision. Science relates all produced knowledges to an object it never "decided" or "autoposited"; and the object precedes, in the order of the given, these philosophical operations.

In effect, this "objectivity" or this "nonthetic" transcendence in general is one of science's phenomenal givens, which philosophy can only deny by attempting to divide and thus to redouble it. In philosophy the "object" is not only given in this quadruple characterization vis-à-vis the knowledge drawn from it. It is also given in its redoubling, in the autoposition of transcendence and thus the autoposition of traits that this transcendence can ground. The "fact" of science, what the philosopher calls "fact" by fixing and reifying scientific labor, is this circular redoubling of transcendence. Transcendence is, at any rate, necessary to the scientific object, but perhaps not in this decisional and autopositional form. Rather, the object must be radically or noncircularly "caused"; it is not caused by itself, but by nonthetic Identity (of) self, which is the cause (of) science and its transcendental determination. Given its immanence without blending, this Identity functions like "Occam's razor" for the philosophical Decision in general and for transcendence in particular. It "simplifies" transcendence by prohibiting its circular redoubling, its vicious autoposition, by destroying at the root of transcendence what constituted its philosophical specificity or use: its nature as decision and as autoposition, which is in some sense, as we will incessantly see, "reduced" by the One.

To manifest its absolute reality, i.e., to recognize a transcendental cause that would not be imposed on it by philosophy, by its project and its will, science (as "subject" of its own description) proceeds not to one but to *two* transcendental reductions. These reductions are clearly heterogeneous by their source, their procedure, and their scope; the second has an immanent source and has already been carried out or completed. If we assume this slightly external perspective, then no less than these two reductions are necessary to "access" this essence. They bear on science's supposedly "crudest" objectivity and realism. Let's assume with philosophy that science is thought's Other, that it is an almost total and "ontic" obscurity, an absolute Other or an Unconscious of thought. Let's suppose with philosophy, for a moment, this point of departure. Two reductions are possible:

1/ A reduction that is already called "transcendental," but is idealist and thus philosophical, resting on an operation of transcendence toward science's ideal(-real) essence. It allows us to pass from the supposedly "in-itself" object to the object as objectivity and as objectivation; from the supposedly transcendent real to the ideal objectivity of the sense-of-object. This reduction is grounded in philosophy or in transcendence; it is precisely a philosophical or epistemological Decision, which forms a system with the assumption of the primary and unconscious realism that, according to philosophy, would be that of science.

2/ A whole other reduction, since it now bears on what subsists from philosophical transcendence in the previous operation. This reduction can thus no longer be philosophical: it is scientific, carried out from the start and already completed by science and its transcendental cause; nor idealist: it is included in the vision-in-One or in the reality of immanent givens that form the absolute determination of science. It suspends not objectivity itself (there is no scientific representation without objectivity), *but its circular redoubling in itself, which is peculiar to philosophy*. Philosophy is not only a practice of the object's objectivity; it is equally its alleged theory in the form of its autoposition.

Science thus forces us to take the concept of objectivity in two different senses. Sciences have a use of objectivity, but objectivity is not autonomous. It is transcendentally "caused" and simplified, not redoubled. It constitutes the nonthetic "real object" (of) self or the a priori to which representations are related. Philosophy, in contrast, is the vicious redoubling of objectivity, which claims to found itself and thus becomes this alleged "rational fact" at the foundation of epistemologies. Science "lacks" self-reflection; it in no way lacks transcendental essence and transcendental autonomy. Science is a simplification of objectivity, reduced to its minimal aprioric givens, and a subjection of these givens to the real, i.e., to the transcendental Identity. It implies the destruction of the illusory autonomy of philosophical objectivity. These results will prove to be fundamental when we undertake the real critique of *alterity*, of the concept of the *Other* that contemporary philosophies assume without elucidating.

THE OBJECT OF KNOWLEDGE

A third, ultimate, or transcendental condition has to be finally fulfilled for an autonomous thought of scientific knowledge to exist: this knowledge must represent the "real object" in an "object of knowledge" (finite product of the process), which has *the property of modifying itself without thereby claiming to modify the known real, as is the case in philosophy*. Knowledge is not an attribute or a determination of the known real, which is a cause absolutely anterior to its knowledge. Science changes the order of its thoughts without also claiming to change the order of the real. It is indeed a representation, but is also nonthetic (of) the real. It is knowledge of the real, but paradoxically it can only be so in a mode that, for instance, excludes the causality through which (philosophical) objectivation claims to modify the object or the being of being. It is thus deprived of every efficacy over the One and over the real object, at least the efficacy that would pass through autopositional transcendence, through the scission "in view of" the position of this object. The distinction between two "objects" is a duality-without-synthesis, without a superior unity, because it does not derive from a scission or a decision. But it is not without cause in Identity or without *occasion* in experience. We can now have a better understanding of the nature of this Identity-without-identification, of the fact that it is not a superior synthetic Unity—since it tolerates and conditions a radical duality, which is nevertheless not without internal relation—but rather a relation of unilateral determination, as we call it, and it goes irreversibly from the real to its representation. Science thus knows itself as science (of) the real, but strictly ordered to the real. This knowing is not grounded in the suspension of every relation of ontological causality between representation and the real; it is founded by what causes this suspension: the specific Identity of the real itself.

Science is then an intrinsically "dual" or "dualitary" activity, traversed by an irreversible cleavage, which no longer results from a scission of a prior Unity, a scission that would necessarily belong to it like the Dyad to the One.

a/ It knows itself as science (of) the real in-the-last-instance and not as science of the "properties" of being for which philosophy alone could supply being, presence, or objectivity. There is a real in itself, which is neither

Being nor being, neither the objective nor its properties nor their "difference." Science does not leave, it has already left the equation of metaphysics (because it never entered into this equation): logos = being, thinking =
the real, which always presupposes a transcendence of Being or the real
to thought, and vice versa, and their identity as the Same. Science immediately inhabits the real; it knows the real and can assert this knowledge
against philosophy. Its essence is a transcendental experience whose secret
is not contained in philosophy.

b/ Although it knows straightaway that it inhabits the real and does not
have to identify with it, to posit it, science knows at the same time that it
is just a simple representation of the real, a simple reflection that does not
constitute or determine it, as philosophy always believes it can. It is thus a
nonthetic reflection (of) the rigorously unconstituted real. This status of the
absolute or mirror-less reflection explains the necessity of a permanent "rectification" of concepts at the same time as their "adequation," which corresponds, if not to a "local," at least to a "finite" and of-the-final-instance
realism of representations. This adequation is thus not the metaphysical
certainty that philosophers, dazzled by their own illusions, want to inject
into science. That science knows that it is a nonobjectivating representation (of) the real does not prevent it—quite the contrary—from rectifying
its representations, but science does so *on the basis* of its subjection and its
adequation. For the scientific posture consists in "assuming"—in a nothing-
but-immanent way—a nonobjectivating Identity-without-identification
and without-synthesis of the real and of representation and, only *thereby*,
in representing it without claiming to transform it in this operation. It is
not a matter of a "poor" adequation: knowledge is not subject to a real as
already represented or given in the mode of autoposition, i.e., to its anterior
representations. But it is subject to the real that is to be known and that is
necessarily already given in-the-last-instance alone and prior to every synthesis. Hence what must be called, in opposition to philosophy's unlimited
journey, science's intrinsic or postural finitude, which forever prohibits its
secession from the real to be known, but not from its current representation.
The representation is therefore voided of every ontological function. And
this explains its reduction to the state of description and the destruction of
its constitutive claims, at the same time that it goes from the circle of autoposition to the state of nonthetic reflection or representation. The scientific use—the one we make here—of language in general and of "categories"

is thoroughly heterogeneous to philosophy's use of "logos," i.e., as constituting or unveiling = realizing the real. Science implicates a descriptive and no longer constitutive use of representation in general, which alters nothing by manifesting it. Whereas science alters the order of its representations rather than the order of the real, philosophy *claims* to alter the latter through the former. Hence its transcendental illusion. Only science can then found another, nonillusory use of philosophy: as nonthetic representation (of) the real. This scientific use of philosophy receives the name of "non-philosophy."

We thus dissolve, through this theory of nonthetic reflection, the philosophical reduction of science, i.e., the thesis according to which science would rest on objectivation. Since objectivation is the theme of philosophy par excellence, science is stripped of its own essence from the start and receives a substitute essence, that of philosophy, of which it is then the degradation. This logic is irrefutable, but it is possible to extirpate it, at its root, from scientific thought.

What happens? For philosophy, it is fundamental to circularly conflate—even if it is in the more or less long run and with more or less delay—the known real with its objectivation, the real and its knowledge, the real object and the object of knowledges, which are supposed to be determined reciprocally. In their hegemonic will over the sciences, philosophers *confuse two heterogeneous modes of phenomenalization of the real*: the philosophical, which implicates Decision or Transcendence as its major operation, and the scientific that excludes such a Decision from its essence or that phenomenalizes the real as already-Phenomenon and keeps it in its most realist and immediate "naïveté," in its immanence, which is deprived of any exteriority. This confusion is naturally followed by another: every knowing is finally reduced to a historical knowing, i.e., to the deployment of a transcendence. In contrast, science in its relation to the real to be known does not proceed through objectivation, which is always a supposedly primary and constitutive exteriority or transcendence. Philosophy is the circular theory and practice of objectivity and of its essence, which reciprocally determine each other. Yet science does not ultimately have an object in this sense or in the *ob-jective* sense. The known real and the knowledge of this real form an irreversible duality. Knowledge is a reflection (of) the real, but a reflection that neither posits nor objectivates it and that has its ultimate nonsynthetic cause in the real. This presumes that science, in its specificity, is *a knowing* in which the observer can modify the objective phenomena without modifying the real implicated *in* and *through* the immanence of the scientific posture. Science does not form a circle with the real as does philosophy, which claims to be not only its knowledge but its coproduction. What is in fact called "ob-jectivity" in general is a confusion of the known real with the (ever modifiable) object of knowledge. Hence the following distinction: philosophy is the theory and practice of the ob-ject or confounds the real with its representation; science has a real "object," i.e., it does not have objects in the philosophical sense; it entertains with the real a "relation" that is no longer one of *objectivation*. There are two heterogeneous modes of phenomenalization that philosophy—always unitary—seeks to conflate, while science seeks the autonomy of its way of thinking the real and the distinction between these two modes.

Science possesses and exhibits from itself an absolute cause of-the-last-instance, an untransformable Identity (of) self and (of) its nonthetic representation. Scientific posture is already complete or finished; it claims—an immanent claim—to have access (to) the real from the start—not to the object of knowledge . . . —without having first to deny or posit it. Truth be told, science posits neither the real it represents nor the representation it gives of the real. Its realism is not secondary, transcendent, or willed; it is primitive or "first."

EMPIRICAL SCIENCE AND TRANSCENDENTAL SCIENCE

In the division of intellectual labor it imposes on science, philosophy draws a final distinction that flows from the others and sums them up: it distinguishes "empirical" sciences and "transcendental" science, that is, itself as science of Being or of Logos. To be sure, this distinction and this correlation, this hierarchy that carries the "empirico-transcendental doublet," and this circle are broken by science itself when it undertakes to describe its real essence.

1/ Every science, even "empirical" science—or a science that philosophy calls "empirical" in order to degrade it—is in reality also "transcendental": it bears on the real itself and, moreover, it knows that it is related to the real. Undoubtedly, philosophy also claims to be related to the real. But this is an abstract resemblance and a generality. For science is related (to) the real itself and knows that it is related to the real twice in a nondecisional and nonpositional way, within the intrinsically finite limits of absolute immanent givens or of the One as Identity (of) self (and) (of) knowing (of) self. It manifests itself as more primitive and more elementary than philosophy and dismisses its claim.

2/ A transcendental science—for example, the science (of) the essence of science sketched out here—is necessarily also "empirical," but in the new sense that it has to receive its object (being unable to create it), taken from matter or from thought, that it obtains this object in the "World": for example, in this case, existing sciences, epistemologically blended with the philosophical Decision, in view of constituting a science that would take this blend as its object-material and would have to receive it without being able to create it. The relation of the aprioric structures of every possible science to the "empirical" object should be clarified more fully by means of the description of the "dualitary order," of the "unilateral order," or of the order of "determination-in-the-last-instance" that regulates this relation.

Although every science participates in the transcendental structures of the One, of the real Object and of the Object of knowledge, we will more particularly call "first science" or "transcendental science"—at the risk of recreating a misunderstanding akin to the old philosophical hierarchy—the particular "empirical" science constituted as "auto"description of the essence of sciences and that must necessarily take the philosophical Decision itself as its object-phenomenon. This will be a "transcendental" science—a first time in the sense in which from now on any science, "empirical" by its object or its worldly materials, deserves to be transcendental by its essence and a second time in the sense in which it is constituted by describing the essence of any worldly empirical science. Yet these two uses of the "transcendental" are, in reality, perfectly identical. Such a science remains empirical by its materials (the epistemological Decisions), even if it is transcendental by the way in which it is constituted (the description of every science's aprioric structure).

In relation to the *circle*, to the divided *unity* of the empirical and the transcendental, in relation to their philosophical hierarchy, which is the hierarchy of a division of labor, the new distribution of the empirical and of the transcendental reflects the "dual" or "dualitary" spirit of science, namely the fact that every science immanently requires both: 1/ an empirical given it does not create but discovers in the World or Effectivity; thus the contingency and transcendence of this given that serves it as materials and as occasional cause, but a cause it cannot produce; 2/ the use of this given in terms of the four scientific a prioris, rather than its presumed objectivation and "categorial" transformation, as soon as it is inserted or included in the form of the "real object," i.e., of the "nonthetic" or non"autopositional""objectivity."

The transcendental (the cause of sciences) and the empirical (the given or occasional materials) no longer form the circle of a divided Unity or a *philosophical Decision*. They at last shatter the famous empirico-transcendental parallelism by distributing themselves according to the relation of the "dual" (rather than of the philosophical unitary duality) or of Determination-in-the-last-instance. We will have to clarify this relation by examining the case of the relation between "transcendental science" or "nonthetic science (of) sciences" and its object-materials: the philosophical Decision.

At the foundation of this new nonunitary distribution of the empirical and the transcendental clearly lies a new experience of the transcendental: within the limits of scientific finitude. Every science indicates itself and starts from itself without having to leave itself. It is not philosophy but, rather, science that is the "transcendental subject." But on the imperative condition that it modifies our concept and our experience of this subject: as nothing-but-immanence and without transcendence, thus as nonepistemological or non-Kantian. It is not the immanence of the reflexive process *of* the Idea of the Idea (Spinoza). It excludes every transcendence, intentionality, or reflection, every logos in general. Science supplies its criterion of reality and of validity from its own foundation. It exhibits on its own its ultimate givens without resorting to the mediation of philosophical operations. It is capable of bracketing the philosophical positions (ontology, epistemology, idealism, materialism, positivism, skepticism, etc.) and philosophical positionality in general.

Thus philosophy has never *really* founded science. It has instead projected a possible image on it, the possibility or the project of a science. Science, nevertheless, does not *found* itself. If it is capable of describing itself, without constitutively resorting to philosophy's technologies, it can do so through the immanence of its cause. In other words, the real content of what we have called the *epistemic* is no longer a philosophy-of-science, an epistemology. It is *a first science, a transcendental science* of a new type. We have—rather, we are, and as *humans*—an absolute, if not effective experience (of) science, absolute albeit naive and unreflected, but quite sufficient to allow us to describe science's internal structure. We can thus pass directly from sciences to transcendental science, which is the theory of their essence, and circumvent the allegedly "uncircumventable" philosophical Decision. Through the One, we pass from empirical sciences to transcendental science, in which empirical sciences describe themselves and retain their naïveté without passing through the philosophical operation as constitutive.

By its transcendental kernel, science can assume some of the functions of philosophy, but it transforms them without resorting to the Decision's fundamental procedures. It does not occupy the place vacated by philosophy. On one hand, philosophy never leaves any place vacant; on the other, a transcendental science is not a philosophical *image* of science like positivisms, which are nothing more than philosophy's autodenials via scientific intermediaries. This is why the reevaluation of sciences against their traditional image programs the necessity of a science of philosophy and of epistemology and can found this project.

What we have described is the real infrastructure, real in the rigorous sense of every science and not in the empirico-logical sense.

To conclude: *we have not described here the scientific labor grasped in exteriority or in the epistemological prism, but science's immanent thought, which gives experimentation and theorization their scientific values, i.e., their relation to the real.* Science is not any combination of more or less already elaborated givens, already constituted knowledges. It only acquires its supposed theoretical meaning under certain conditions—the One as determination-of-the-last-instance, with the "realism" that follows from it—and these conditions are spontaneously realized in the scientific posture. If there is a possible description of science, this description requires us to take science right at the level of its phenomenality, insofar as it is nothing other than *immanent phenomenon* (of theory, of givens, of experimentation, of deduction, etc.). We thus take it outside transcendent constructions and philosophical interpretations (conditions of possibility, dialectic, hermeneutics, structuration, etc.), which are spontaneous but invert science's meaning by projecting it in the wholly other space of reflection, of consciousness, of the concept, and of the philosophical circle in general. All these elements intervene locally in the complex scientific practice, but certainly not in its essence. And a "first science" must keep them absolutely in abeyance.

As it will be noted, as it will perhaps be objected, we have refrained from supplying an epistemological description. This is not our problem. Instead, the problem to which we should draw the attention of scientists, and perhaps of philosophers as well, is that it is less urgent to fabricate a new ad hoc epistemology on the transcendent presupposition that science is an empirico-rational fact than to know if *epistemology* in general has some meaning (and what meaning) for science. Before being an alleged empirico-rational and fetishized fact, science is an immanent infrastructure. Our point of view—but we cannot demonstrate this in detail here—forces us to posit the scientific impertinence of epistemologies (impertinence at least as to the essence of science). And then to posit the equivalence—with respect to science and from its perspective alone—of all the possible epistemological Decisions. The Principle of equivalence of all the epistemological observers of science cannot be in its turn a philosophical Decision; it flows from science itself when science is restored to its essence and when this essence shows itself indifferent to philosophy.[1]

2

NON-PHILOSOPHY

A SCIENTIFIC REFORM OF THE UNDERSTANDING

"NON-PHILOSOPHY" or "first science" seeks to be "scientific" in its method and its essence, but immediately touches on philosophy in its object. It strives to extend the criteria of scientific thought, which is rethought beforehand in its autonomy, to the theory and practice of philosophy. But it is not in any way a mode of "philosophy as rigorous science" or the negation of philosophy. It is a positive, autonomous discipline, which aims to inscribe itself in the lineage of sciences while rectifying their epistemological concept. It proceeds through operations of the type "experimentation" and "deduction." But, on one hand, it remains transcendental and nonpositivist (hence its other possible designations: "transcendental science" or "first science") and, on the other, it applies its operations to philosophical statements, thus to "natural" and not logically formalized language.

Let's place this entire project under the rubric of a *scientific reform of the understanding*. A reform of the scientific understanding would have a philosophical inspiration. This is not the case here where theoretical thought, the "understanding," ceases to be copied (as in the theories of knowledge and philosophies in general) from more or less sublimated, interiorized, or displaced *faculties*. The knowledge at issue is the one engendered by the scientific process, whose structures and aprioric operations have nothing in common with the faculties. Theory of science rather than theory of knowledge: such is the content of "first science," or again: theory of scientific-knowledge-as-thought rather than philosophical-thought-of-knowledge.

A few major theses or, rather, a few *axioms*, sum up non-philosophy. Their exposition will clarify the descriptions of the previous chapter and introduce new supplementary clarifications from the perspective of a "first science" and of its use of philosophy.

THE ESSENCE OF THE REAL RESIDES NEITHER IN BEING NOR IN THE OTHER, BUT IN THE ONE

This—real—axiom has to give us the means to rethink the One theoretically from itself as cause or "qua One" and outside every convertibility with Being or the Other. It has to give us the means to subvert—at least in this science itself—the founding presuppositions of metaphysics and of its deconstructions, i.e., philosophy's entire authority over the real. This thought of the One, insofar as the One is absolutely "outside-Being" or identical to the real itself, forms a "transcendental science" because it discovers in the One *its cause* and not only, like philosophy, a theme or one of its objects.

A thought of the One is devoted to the real alone, and the real should not be confused with the Whole or Being. But we will not say that it is abstract or too exiguous. This would amount to confusing the One qua One, defined by a radical and thus nonpositional and nondecisional immanence (of) self, with an ontological One, pervaded by transcendence, impotent outside its convertibility with Being, a One "without-Being" in the sense in which it said of *Dasein* that it can be "without-World" (*Weltlos*). Science is certainly not a super-, a meta- or an inframetaphysics. In science the One is not torn away from its element, and everything else negated, excluded, as it is in Parmenides' One-All. Not only does it let Being and being subsist "outside" it; but, as cause (of) science, the One is vision-in-One, seen-in-One of the World, of Being and of being to which it unconditionally gives phenomenalized being. It is also outside every metaphysical parousia. And for that reason it includes the World, Being and being in the immanence of the theoretical space, a space rooted in the One itself and received or given, manifested in the mode peculiar to the One. Nothing of what is not in the

One, nothing of Transcendence is negated or destroyed by this phenome-nalization, if not the philosophical Autoposition. It ceases only to be a point of view on itself, a support of a philosophical thought that always mani-fests by claiming to transform. It is important to distinguish (a) a thought founded on the criterion of radical immanence, which only negates the illu-sion, but offers Being its own "phenomenological" generosity or its power of manifestation, and (b) a thought founded on the permanent autosurpassing of transcendence, which leads to abstractions and self-attenuation.

NON-PHILOSOPHY'S OBJECT IS PRIMARILY PHILOSOPHY, THEN, THROUGH PHILOSOPHY, ALL THE OTHER POSSIBLE STATEMENTS; IT IS THROUGH THIS OBJECT THAT NON-PHILOSOPHY DISTINGUISHES ITSELF FROM OTHER SCIENCES

Non-philosophy is not "on the margin" of philosophy. Rather, it is philoso-phy that ceases to be the site of non-philosophy or its foundation and is reduced to the state of materials or of "objective givens," "phenomena" that can be treated by the operations of this science. Thus philosophy also cannot be the *object* of science in the philosophical sense of the term *object;* it is so in the precisely scientific sense of this word. This sense will become explicit through the forthcoming description of the structures of scientific theory's or knowledge's fractality.

In general, a science is designated—this is a tradition—by its "objects," the region of its phenomena. This is equally the case for the science *of the One.* This formula does not mean that it is science of the One to the exclu-sion of philosophy (cf. the possible hesitation: "science of the One" or "science of philosophy"?); rather, it is a science that makes use of these phe-nomena that are *the philosophical statements about the One.* Since sciences' cause is equally "qua One" (the One as absolutely given), "science of the One" does not always designate this One-cause, but also the philosophical or phenomenon-One. So there is no contradiction between the two formu-las: science of the One or of philosophy, because the One at issue in these formulas is either its material, that which is provided by philosophical state-ments, or its cause, but without any amphibology between them. From the first perspective, the science of the One is distinguished from other sciences precisely because it is a science of philosophy, insofar as philosophy aims at the One-of-which-it-speaks. From the second perspective, the guiding formula can simply signify the real state of affairs of the One-cause, if it is understood as follows: *the being-One (of) the One, the being-Identity (of) Identity*. Philosophy can at most lead only to the *being* of the One or the *being* of Identity and bars the One by Being, which represses it.

By its essence or its cause, we will call this discipline *transcendental* or *first science*; by its material (philosophy) and its product (non-philosophy in the strict sense of the term), we will instead and generally call it *non-philosophy*. This term and its "non-Euclidean" analogy make it sufficiently clear that it is not a matter of negating philosophy or "leaving" it—impossible operations—but of recognizing that its alleged validity and its claims over the real have always been suspended by science. Furthermore, such a science would have no sense, at least as a theory, if it did not let itself be determined—within some precise limits—by philosophy as well as by its material. It entertains with this material a *scientific* rather than philosophical relation or a relation of objectivation; a purely transcendental, rather than empirical or mixed relation ("empirico-transcendental," which is philosophy itself).

Thus the guiding formula (the One qua One, Identity qua Identity, and so forth) no longer has the ontological sense of tautology, of "as such" [*comme tel*]. The One intervenes twice, but in a mode that we will henceforth call "dualized" or "unilateralized." The One as cause (of) science, as absolute Given, conditions the transformation of philosophical statements and their insertion into a truly theoretical space. In this way, the unity, the convertibility of philosophy (thus the convertibility of Being) and of the One itself is broken a priori.

NON-PHILOSOPHY IS PRACTICED IN THE MANNER OF A THEORY

"Between" its cause and its object, between its essence and its material, or rather from one to the other, it produces *theoretical* and no longer *philosophical* thought: the relation of its cause and its object excludes the conditions of philosophy. Against the spontaneous confusion of philosophy with culture, against its inevitable "cultural" and "ideological" drift, non-philosophy aims to safeguard the rights of theory and to struggle against the more initial and foundational confusion of science with philosophy. It thus struggles against the "philosophies of science" and epistemologies. Here theory no longer has a philosophical essence because philosophy was never a theory, a knowledge that is grounded in reality and in rigor. First science enjoys a universality "superior" to the one resulting from the distinction and combination of "generality" and "totality" (the two sides of onto-theo-logy). And it introduces a mutation or a generalization of the non-Euclidean *type* into the practice of thought in general. It is to philosophy what non-Euclidean geometries are to Euclidean ones within the "genus" of geometry. Moreover, it is to philosophy what a science is to a nonscience.

These first axioms are not enough to make non-philosophy concretely intelligible. They have to be completed with others (always transcendental and not logical) that determine the field and style of non-philosophical practice.

PHILOSOPHY DOES NOT REACH OR KNOW THE REAL

What the critiques (Kant) or deconstructions (Heidegger) of metaphysics demonstrated about the extent of the transcendental illusion or of its errancy is still limited. And these authors did not draw all the consequences of their description of the structure of the philosophical decision. Globally, it is philosophy as *decision* (including the one that pursues the—strictly partial—deconstruction of the decision) that must be deemed the victim of an *appearance*, i.e., of a "forgetting" of the real, more profound, more extensive than the simple forgetting of "Being-as-forgetting." *After* the analyses of the philosophical gesture, of onto-theo-logy, of presence and of decision carried out by Heidegger, Wittgenstein, and Derrida, it seems hardly possible to try once again, like these authors, to make sense of things and recognize to philosophy an ultimate right, the right of its claim to codetermine the real. It is possible only to extend the transcendental illusion of the *dialectic* of (dogmatic and skeptical) metaphysics to every *decision-* or thought-by-*dyad* (mixture, blending, doublet, hierarchy), as is philosophy in its most secret operations. It is in effect a *transcendental dialectic* in the sense that we can generalize of the term *dialectic:* not only in its rationalist-empiricist form or in its onto-theo-logical form (the *difference* of horizontal Being and Totality, of ontology and theology), but in its more general form as *decision*, dyad, or unity-of-opposites, with the operations that animate them (reversal and displacement). Contemporaries have shown the depth of these schemas, deeper than their dogmatic version, so long as they are divested of their restrained rational form. And the true "return to Kant," if it is not a simple return to the doctrinal and obsolete Kantianism, means that science alone—as thought of the real itself—can demonstrate this universal transcendental illusion.

Thus philosophy does not reach the real in its identity but reaches Being, i.e., a blending, a combination of reality and of appearances or simulacra. Being's semiunreal constitution, its content of appearance or of imagination (*ens imaginarium*) proves that it is not limited to the nothing-but-real, but that it seeks to be itself through this latter and the functional use it makes of it. It extracts from it a surplus value of unreality, possibility, image, and transcendence from which it draws its prestige and, in this way, asserts that it is necessary to the real. We must limit philosophy in order to make way for science—on the condition, perhaps, that from now on it be up to science to make a place for philosophy . . .

THE REAL DOES NOT RESIDE IN BEING, BUT IN THE ONE-WITHOUT-BEING, AND YET IT IS THINKABLE: IT IS "DETERMINATION-IN-THE-LAST-INSTANCE"

The real in the irrefutable sense of Given-in-immanence anterior to every operation (even the operation of donation) is the One, but "qua One" or as "last instance."

Being testifies only to philosophical hubris, whereas the One is the nothing-but-Given that precedes every donation. So much so that the veritable order of transcendental reasons goes from the One to philosophy and not from philosophy to the One. This order upsets all the traditional—globally idealist—relations between the real and its thought. The thought of the real must be able to think itself and to expound itself at once: 1/ as a thought that stands in strict dependence on the real-One, that does not coconstitute it as it coconstitutes Being, but receives the real-One as the given absolutely anterior to it; 2/ as nevertheless capable—despite its pos-teriority—of thinking the One, of constituting the *adequate* representation (of) the real, and thus by definition it never objectivates the real in its rela-tion to it. What concept should we introduce here? The true relation of the real-One and of thought, if it is *no longer in any way* a (total or partial) objectivation of the real-One by thought, can only be formulated as *deter-mination-in-last-instance* of thought by the real-One. This is no doubt the most profound sense of the originally Marxist formula.

Parmenides' guiding formula: "The Same is Being and Thinking," which structures every philosophy and even partially or half of its deconstructions, is eradicated by the One itself. As soon as it is a matter of the One that is thought without-Being or outside-Being—a possible thought, as we have just seen, if the traditional and philosophical relation of objectivation and reciprocity is replaced with the determinant-in-the-last-instance cause—"real and thought" are no longer "the same." Nor are they simply "identical": *real and thought are identical in-the-last-instance alone.* The One is the last—or first, but absolutely first—instance as well as the transcendental instance that "holds" within itself (but without being affected by it) the thought that will represent it, that will even be its *adequate* albeit belated representation. More than belated: de jure "posterior" or lagging in relation to the One. More profoundly: uni-lateralized by it.

Non-philosophy has no other real beginning, if not no other founda-tion, than the rectification, in a scientific and axiomatic sense, of our onto-logical representations of the One (its convertibility with Being or with the Other and its consequences). This rectification represents it now as preced-ing Being (and thought) in the *ordo essendi* as well as in the *ordo cognoscendi*. More profoundly, it begins with assuming this order as the only real one, thus with the One as the *Given* itself, the sole given that is given indepen-dently of every donation.

A THEORY OF THE ONE—OF IDENTITY QUA IDENTITY—IS NECESSARY, AND IT IS ALREADY GIVEN BY SCIENCE, PROVIDED WE KNOW HOW TO THINK THIS SCIENCE

A different global schema of science than the one offered by philosophy and epistemologies is possible and necessary. A science's non-philosophical economy is the following: qua theory, it is related in-the-last-instance alone, or in the above-indicated mode, to the real-One, to the nondecisional Identity (of) self as well as to its cause; and it is related, on the other hand, to its "objective givens" or the "phenomena" of its domain in a mode that is not that of philosophical objectivation, but the one we call the mode of "materials." This mode is specified within the functions of index, material, and symbolic support. Every decisional and positional (= objectivating) schema, its modes and its temperaments (intentionality, ekstasis, project, horizontal opening, etc.) must be eradicated from the relation of theory to the real or to the cause as well as its relation to "phenomena." Not only are they no longer conflated and no longer communicate as in philosophy, but the real-One is what unilateralizes—displaces without return or reciprocity—the "phenomena" or *data* and thus what frees up a specific place for theory or thought, free finally of every confusion with the "empirical" givens. Determination-in-the-last-instance "and" materials form a system, and both derive from the cause. This new architecture is probably the real economy of science at work, the way science can think itself outside philosophy's codes, injunctions, and norms. As the real-One is "without-Being," science is "without-philosophy"; the "without" does not indicate here radical absence or negation, but a change in use and the abandonment of their function, which is constitutive of the real itself.

To understand the type of displacement to which we are subjecting the philosophical images of science, we can recall that at first these images invariantly set in play a duality (concept and experience, operation and object, knowledge and data, logic and fact), then a superior unity (dialectic of concept and experience, reflexive or operative unity, objective transcendental, research programs, etc.). It is this overall schema that is overturned, disorganized, more-than-displaced by "science itself" or qua science that thinks itself. Instead of projecting a superior unity of synthesis, which gives scientific labor a circular and thus philosophical appearance, science works unilaterally, *without synthesis or circle*. It prohibits every "hermeneutic" or stylistically philosophical reappropriation, which reaps the fruits of scientific labor. As to this labor, it is enough to substitute Identity for Unity and Determination-in-the-last-instance for synthesis and reciprocal action (dialectical or otherwise) so as to autonomize theory and experience in relation to each other and render theory truly falsifiable in its restrained particularity, to no longer confuse the movement of knowledge and philosophical "becoming," which turns endlessly in the doublets. No term is any longer in a "face to face" with the other or forms a fold with it; all terms are distributed on the axis of asymmetry or of unilateralization. And yet, in each of these stages (Identity-cause, theory, experience and its symbolic use), it is indeed Identity that reigns: either alone (real-cause) or as last-instance of theory, of theoretical immanence and of the manifold of determinations, or, finally, as "symbolic" identity that can be the identity of experience and its givens, reduced to the state of "symbolic support" (we do not describe here the details of scientific labor). So much so that, ceasing to be in a doublet or in a face to face and thus to be sterile, the concept and experience, theory and the theorico-experimental givens are now only *related* to one another from the last-instance alone (something that frees them in their own positivity) and not in the indeterminate generality that is philosophers' "superior Unity" (dialectical, reflexive, transcendental, operative, etc.). Science's general economy, its organization, does not rests on the structures or principles of philosophy (Dyad Unity), but on ones peculiar to it (Identity → unilateral Duality). Science is thus a *unified* thought and not *unitary* like philosophy; it is also a simpler thought than philosophy—we will come back to this.

SCIENCE AS POOR OR MINIMAL THOUGHT: OCCAM'S RAZOR

Science as practice of thought is the critique in action of philosophical operations. These operations are founded on the reciprocity of two terms: amphibologies, analyses and syntheses, diverse or structured dialectics, and so on. It opposes to this active and interventionist style a style of pure description, absolutely contemplative of Identities, i.e., of phenomena as such or of nonmodifiable phenomena, of terms and their priority over relations, of the identity of orders or instances, and so on. Each term, word, or category that appears in the space of science must be described as though it constituted an autonomous order. Even the pure, infinitely open rapport, the order of theory in the narrow sense, forms a relatively autonomous sphere that can be described without any supplements. This is due to the Identity that is required here as a transcendental guide, *last-instance that applies to all orders*, even the most complex. So much so that in order to be an autonomous identity, a scientific object (a scientific knowledge) is not an abstract identity outside philosophical mixture and dependent on it—since the mixture itself, the philosophical, is in its turn defined by its identity-of-the-last-instance.

Identity operates like a prodigious Occam's razor. But in the name of a real simplicity and of a real poverty of thought, not in the name of a supplementary simplification or impoverishment of philosophy. Not against philosophical or metaphysical entities that are judged to be useless or cumbersome; but for itself as science, that is, "against" the fundamental philosophical disposition, against spontaneous philosophy, which is the *autoposition* or the *fold* of every transcendence left to itself. To philosophize is to trace a more or less split, partially open circle around a category, a concept, or an experience drawn from transcendence in general: of the World, of History, of Desire, of Power, of Perception, of Language. To philosophize is to start the circle from this supposedly already given entity, to make it traverse and split this entity, to displace and reunify a little farther, for no gain *other than this operation*, no gain as to the real itself, i.e., to the Identity (of) this entity. To philosophize is to posit oneself by repositing oneself, give or take a scission or a decision that does not undermine the real and produces the flocculation of the Same. To philosophize is to require every thing, and thought itself, twice instead of once, and twice in such a way that none of the strokes or their sum produces something other than itself, in other words: an absolute inequality to the real.

Science thinks the real all-at-once, without dividing it and without dividing itself. This is why it thinks a multiplicity of each-time-one-time, a veritable multiplicity of undivided "terms" or a chaos. That philosophy thinks twice does not mean for science that it thinks once too many, but that it is this 2 of the division, i.e., this 1 → 2 or this 2 → 1 that is too many, useless and uncertain. In wanting to intervene half a time as transcendent object = x, half a time as this transcendence, and finally as the unitary synthesis of the object and of its objectivation, it is globally philosophy (rather than one of its operations) that is in excess.

Identity, as it is implicated in science, is precisely the real whose essence is deprived of this structure of autoposition or of folding. Strictly immanent, thus immanent only one time and without complexity, timely and without delay, it renders all that folds or refolds, all that is announced in the panorama of Transcendence, necessarily contingent. When it has a scientific origin, Occam's razor suspends philosophy globally, its very claim, instead of some superfluous notion. It suspends philosophy in an unappealable contingency, sterilizes it in some sense without destroying it. Identity then has the World and philosophy as its correlate; from now on, they are—for transcendental reasons—simple, quasi-material, nonsignifying and inert residues. They should fulfill other functions—in science—than those spontaneously assigned to them.

Science is in essence a simple and minimal thought. This is not the result of a philosophical reduction or a metaphysical ascesis: a simplicity of essence, but not of structure, which holds in suspense the complexity of division, of abyss, and of autoreflection that is philosophy's. We know that philosophy's liquidation of back-worlds failed. Philosophy operates globally as the World's form, i.e., as the matrix of every possible back-world. It is in science that this liquidation takes place without remainder, because science is a certain experience of the real as ontologically invisible, but in-the-last-instance alone. It is precisely because Identity is manifest or given through and through that it is invisible to Being.

Nietzsche, Heidegger, Wittgenstein, Derrida, Deleuze . . . equally critiqued *Logos*, *Representation*, but conserved (simply purifying it to various degrees) the matrix of the philosophical, the differe(a)nce or the convertibility of Identity and of the Other. All of them conserved the fold as a last residue of transcendence, of the circle, of the back-world, of autoposition. Singularities, Catastrophes, Multiplicities, Language Games, Dissemination: these are the back-worlds they bequeathed to us. They thought they were liberating us and easing the task, but they simply showed once again the force of the strange enchantment by philosophical faith.

FIRST CHANGING OUR THEORETICAL BASIS
IN ORDER TO CHANGE OUR CATEGORIES

A reform, or better yet a recasting, a non-philosophical redistribution of the traditional economy of thought is needed. *Learning to think* according to science's paradigm rather than according to philosophy's forces us to reformulate new categories, to describe new uses and divisions—to reform the understanding itself through its introduction to the experience of science. We should, however, be more specific.

The shift in the paradigm of thought—but this is a consequence—implies less a shift in the categories than a general rejection of the amphibological confusion of philosophy's and science's categories. To be sure, science no longer proceeds with doublets of the type means/end, agent/effect, matter/idealization, empirical/abstraction, form/information, and so on. These doublets have a philosophical origin, and epistemology projects them imprudently into the sphere of science. Science's major categories are instead: One or cause; last instance; immanent process; material, index, and support; theoretical and symbolic, and so on. We can also add the properly scientific operations (induction/experimentation and deduction) that replace philosophy's operations (reversal and displacement). In particular, science is defined by the "dual" *Identity/Support*, *Last Instance/Occasions*, which do not implicate any decision and thus none of the philosophical hierarchies whose movement takes place through inversion or displacement.

But the true mutation lies in a new use of categories: no longer in terms of their form-mixture in doublet, but in terms of their "vision-in-One." For example, "theory" ceases to be a double or a more or less frozen or deficient mode of philosophy or, in contrast, a "philosophy-as-rigorous-science," in order to become in its turn a pure "phenomenon," an experience manifested-in-the-last-instance in the mode of the One. The description of the phenomenon of theory in general is one of the objects of "first science" as science of the essence of science. It acquires in this science its *identity* of "order" or of "sphere" and ceases to be conflated with a rationalist mode of the philosophical decision or else with a mode of empiricist abstraction—with another philosophy.

To understand thought—the thought of the One—as a science and science as at bottom a thought, in the manner of their "unified," nonunitary or nonhierarchical theory, is certainly to rectify our spontaneous logic, the logic of our relation to the "real," and even to the most obvious phenomena of our ambient world. It is not only a matter of changing our "categories" or even our philosophies, taking up a more current philosophy in place of an older one . . . Nor even changing our alliances, forging new ones in a general strategy of conflicts between science, philosophy and culture, in order to improve our "management" of these conflicts. It is a matter of changing our *manner* of thinking; more than manner (conversion, reversion, trope, etc.), *basis or terrain*: to think *from* Identity rather than *in* Difference (or Being, etc.); *after* Identity or in its descent rather than *by* Difference and in its turnstile; *in-One* rather than *by and for* Being, and so on.

So conceived, through its immanence to the real practice of sciences, the understanding represents in some sense thought's degree 1 ("first")—its minimal form, its primitive and inescapable emergence, the ground that is more solid than every "Being" or every "Foundation." Arche-thought if you wish or the arche-given, from which we could elaborate what technology, ethics, art, love, and politics really are outside their philosophical images, and repose for example the associated problem of "intellectuals" . . .

OF SCIENCE AS MANIFESTATION OF THE REAL

Take two axioms: 1/ the real is conflated with Identity as such or as received and lived by itself; 2/ it is, in this form, science's cause (of) self. These two axioms allow us to unphilosophically pose the problem of knowledge-as-thought and to dissolve some philosophical aporias:

a/ To manifest is no longer to produce, to think is no longer to intervene-in . . .
These confusions must be abandoned. To intervene in . . . the object or to modify . . . the real: these operations prohibit knowledge in the name of interpretation. To think is no longer for example "to reflect," since to reflect is to claim to intervene in the real of thought and to modify it on its own basis and through its autointervention. It is no longer to produce or even to reproduce some real (production and simulation are convertible), but to produce and reproduce only thought in terms of the real alone. It is to manifest the real absolutely as it is without this "as" signifying the "Same" or the philosophical Tautology, which postulates the at least partial identity of the real (as being) and thought (as philosophy). *As* signifies here that the real alone, without thought (as addition, decision, supplement, etc.), is described in the mode of thought alone, which is thus in its own way a specific *order* or a relatively autonomous order. Thinking cannot consist in producing the unproducible Identity; rather, it consists in reproducing thought-knowledge under the conditions of-the-last-instance of Identity. The immanent exercise of the process of knowledge, rather than some of

its operations or its "results," is the manifestation of its cause. Thus thought does not coproduce Identity, but instead manifests it by producing validated knowledge. If the real were knowledge itself, its knowledge would not know it as real, but only as knowledge; and it would involve it in a circle of perdition. Instead, the real is the known, and so the knowledge of the real does not transform it, but ultimately is itself transformed "in view" of the real. Identity *as such*: knowledge thus only manifests the *as such*, but is not confused with it and does not even bring it about. The *as such* instead designates the One-cause that "moves" science. A manifestation of the intrinsic, of the One *qua-One*, is possible, but on the condition that the manifestation no longer claims to intervene in it, to transform or reflect it. It is the very existence of theory as "identical" process that manifests Identity's "interiority," so to speak. But this Identity holds only in-the-last-instance for the theoretical that manifests it.

b/ *To manifest is not to unveil, to think is not to reminisce or to let oneself be overtaken by an anamnesis—nor their opposite or their supplement: to withdraw, to be forgotten.* No more than those of knowledge, even less than them, these operations do not belong to the essence of the real, but to what thought discovers within itself from a more profound *forgetting* of Identity, absolutely consumed forgetting without a possible anamnesis; a forgetting that is otherwise called philosophy. Contemporary thought, inverting and displacing from an "Other" the spontaneous course of metaphysics, believed it could oppose these effects of the Other to the tradition. It merely extracted more crudely the invariants it obeys without much haggling. More profound than the deconstruction of presence is the dissolution of the alleged originary continuity between the real and knowledge, the One and Being, but also the One and the Other. There is a more-than-deconstructive task; it lies in the emancipation of the real's absolutely autonomous order and the relatively autonomous order of its knowledge. Philosophy knows as liberation only the loosening of the knot, the de-stricture, all the operations that prepare other stronger strictures, other more extensive confusions: *more* identification and *less* identity; more closure by dint of reciprocity or of reversibility than opening or unilaterality. Philosophy more or less denudes knowledge in relation to the real, but it does not realize that it is in itself the ultimate, the most irreducible *point de caption* that founds all the aporias and all the illusions.

c/ *To manifest is not to produce some manifest; it is to manifest the already Manifest itself (Identity) and to do so in the "secondary" mode of Manifestation.* Theory is this Manifestation, to be sure, but it does not manifest the real as a being is manifested or as philosophy manifests Being, through objectivation or another mode that always integrates with objectivation (withdrawal, differe(a)nce, etc.). To manifest—in the theoretical mode—the already manifest-real in its own mode, it is enough to produce knowledge. For knowledge is an identical and specific order of reality so long as it serves— by its very existence rather than by aiming for it directly—to manifest the real, not "a second time" but in a whole other mode: as an effect makes visible (insofar as it is an effect) its cause. The One is the only instance that thought cannot aim for, intend, project, and so forth. Theory does not manifest some brute, primary, and generally transcendent real, some real inscribed in the World, even if it is a World of idealities. Nor does it not manifest essences, which are already local crystallizations of knowledge. It manifests the real—which is never a knowledge—by means of the material of essences or of idealities. Science, in the narrow sense of validated theory, is globally an effect—there is an *identity* of the effect—which indicates its cause in its very identity.

"Phenomenon"—what science describes in-the-last-instance by producing knowledge—is defined by its internal or intrinsic identity, by its "static" character. Even the phenomenon of movement (for example the movement of the theoretical process) is "static" and represents a *phenomenal state-of-affairs.* This is how science manifests what is already real as it is, without introducing into it the movement of knowledge. Science does not impart an "ontological" meaning to the movement of knowledges' rectification; it does not postulate that to know a phenomenon is to make it befall its essence. This is the static or descriptive sense of the most mobile knowledge: dynamic in itself, by the knowledges it invalidates or produces, it is static by its effect and above all by its "object." Among other possible objections, philosophy will no doubt say that it is a matter of a formalism, with knowledge indifferently modifying itself without its object being modified. It is, however, only the real (as its cause-of-the-last-instance) that knowledge cannot modify—and the real has nothing to do with an "object." On the other hand, it modifies itself (its "*objects*-of-knowledge," precisely); it can only transform itself. Thus there is authentic knowledge only because the real is not modified with it.

TO REFORM THE UNDERSTANDING ACCORDING TO THE PRINCIPLES OF "FIRST SCIENCE" IS TO PROCEED INDEPENDENTLY OF LOCAL KNOWLEDGES

It is to liberate the understanding from these knowledges and thus from the philosophical closure that, like a parasite, captures them and gets entangled in them. It is to open thought to an absolutely universal dimension, more universal than the time and space we are capable of "thinking" through philosophy or knowing through science. It is to liberate philosophical *ob-jects* or to treat them as simple occasions. Philosophical objects are constructions based on scientific knowledges, artifacts that encumber thought by imposing on it their structure of folding, of autoposition, of doublet, of duplication, or of tautology (the "Same," "Difference"). All of this inhibits or paralyzes thought and imposes on it a teleological closure, more or less "open" but never contested or suspended as such. The entry into theory is the entry into a purely relational, infinitely open space (one necessary term + one radically contingent term), which can be transformed without limits. This space is capable of destroying the philosophical codes or revealing every philosophy as a globally contingent and "imaginary" enterprise of coding. Thought does more than get out into philosophy's sea, its Great Wide Open [*gagner le Grand Large*]—the nearby coast, the prudent coastal navigation and smuggling; it explores the *Wide Open*. We would readily say, if the expression did not lend itself to confusion, that it is *being-in-the-Wide-Open*; *being-in-Horizontality* or *in-the-Spread*, an immemorial farewell to the horizons. This is also what we call a "non-Euclidean" universalization of philosophy.

FROM THIS RESTORATION OF SCIENCE'S ESSENCE FLOWS A NEW DISTRIBUTION OF THE RELATIONS BETWEEN SCIENCE AND PHILOSOPHY AS WELL AS THE PRACTICAL RULES OF NON-PHILOSOPHY

We will no longer examine here the general problems of these relations, but their new de facto distribution: the suspension of philosophy's authority, its constitution into the simple material of a science of the One, which is now possible in accordance with science's new general image. We will indicate, precisely at the interior of this science that uses philosophy as a material, the main rules of the new practice of philosophical statements.

These rules globally transform the decisional and positional structures. They do not derive from the philosophical decision itself, but from the wholly other instance of science: they are the rules of theoretical practice. Instead of difference, dyad, or unity-of-opposites, they emphasize a radically asymmetric or *unilateral duality*, the a priori that defines the space of theory and unconditionally opens the dyad.

On one hand, these rules cannot reprogram mixtures or blendings, the assemblages of the dyad, the economy of its internal operations (reversal and displacement) that decide on the lags, the strategic and reciprocal movements, between coupled and simultaneous terms. They are, on the contrary, a prioris that *assert* the One or Identity in the material and that, by definition, no longer program singularities, catastrophes, or differe(a)nce, but program the effectuation-without-mediation of Identity, the law of radical immanence, in the manifold of the form-dyad itself. This is what we call "vision-in-One"; it is not a transcendent or mystical contemplation of the One, but, if you wish, the immediate implementation of science's operative mystic kernel in the effectivity of philosophy, i.e., of the World.

On the other, they are the rules of science itself. The One "applied" to philosophy's dyad-form produces from it—we cannot demonstrate this here—the de jure non-philosophical, more universal and more powerful theory-form. Theory or science—in the real or the transcendental, not the empiricist and positivist sense—is the perspective of radical Identity or the One-cause on the manifold, *directly on the manifold without any mediation*. It is thus not its blending, its *difference* in the philosophical sense. Theory is in principle falsifiable or surpassable by . . . and in another knowledge; it thus does not include within it any possible, any imaginary, appearance, circularity, or mixture that would in fact prohibit its "falsification" and would allow it to save itself by continually transforming itself. Here again the major principle is to rediscover *Identity*—the Identity of theory or of every other order or instance.

More precisely, what do the dyads or the assemblages, which are philosophy's tissue and "dialectic," become "in-One"? Is it a matter of a snatching away or a subtraction, if not a negation? Rather than reversal, displacement, negation and so forth, the properly scientific operation is globally of the order of *an immediate but of-the-last-instance identification of old contraries, which is not their totalization and which lets them be in their diversity and their identity as terms.* A subtraction would remove a term = X from a dyad or would modify properties of its structure, but it would allow this form itself to subsist intact as the ultimate law of thought. On the other hand, the effect of the unary Identity that acts at the heart of opposites is the extraction and the manifestation of a new instance, irreducible to philosophy's closed dyad. It takes the form of this *uni-lateral duality*, of an original space—the theoretical—which is deployed from the One to the dyad itself. But this dyad is reduced to the state of a henceforth contingent term, without efficacy on the One. A space so finite that it is boundless (neither the One nor the reduced dyad play this role), unlimited like chaos. *Unilateralization* as a generic term designates the ensemble of these effects (effects of theory or of knowledge) that the One-cause produces on the philosophical material and its dyadic structure. An effect of various types according to the "eidetically" distinguished layers in the philosophical decision, which serves us as materials and which we will set out in detail and more precisely elsewhere. It is crucial to grasp that the effect of radical Identity, the identity that is nothing of that on which it acts, is not a mediatized identification, a totalization or other absurd interpretation of this kind. It is: 1/ The immanence of unlimited theoretical transcendence. 2/ A setting-at-a-distance or *in* unlimited distance, without possible reversibility, an absolute loss, the remainderless suspension of every philosophical pertinence. Hence this effect of distancing without return, of "unilateral" separation or dualysis of the philosophical mixture between its philosophical economy (the unity of opposites, their co-belonging), or rather its autopositional pertinence, and all kinds of determinations of this material, including the dyadic co-belonging that has become in its turn a simple "given." 3/ An inscription of this manifold in the immanence of theoretical-being. There is, in fact, a conservation of multiple determinations contained within the philosophical dyad and held in suspense in this aprioric element of theory.

Scientific practice thus consists in describing, with respect to any material whatsoever (in this case, the philosophical material): 1. The uni-lateralizing rejection, the distancing suspension of the philosophical forms or of the anterior and transcendent organization. 2. The effect of nonpositional identification, in-One, of opposed terms or of materials, but from the perspective of these materials, of their transcendence as objective givens or phenomena. The result is what we call science's nonthetic a prioris (we saw that there are four types of them). 3. The effect of this same Identity-of-the-final-instance on the opposed terms—which are conserved "concretely"—but now from the standpoint of theoretical immanence; avoiding the reintroduction of every structure of coupling, every philosophical syntax. The result is an entirely original and concrete "synthesis," the synthesis of theoretical a priori immanence and of its objective givens, but it has now lost every transcendence and passed to the state of *theoretical manifold* or specific manifold: the object-of-knowledge. The same philosophical "couple," the same given material is transformed: its representational content remains invariant, its philosophical syntax is suspended, but this content takes a new original and positive form, that of knowledge.

As to point 2, a supplementary clarification is needed. The theoretical instance does not have a homogeneous nature. It is structured and specified, so much so that the general rule of identification/unilateralization must be diversified and must give place to more precise rules. In effect, the theoretical plane, founded as identity-of-the-last-instance of contraries, represents the level of the a priori, an a priori called "real" in this case. The a priori, if it has its essence in the One, is nevertheless, here as elsewhere, *specified* by the corresponding structural moments of the experience or of the object of which it is the a priori. They are the moments of the philosophical Decision, in which they naturally have the form of the Unity-of-contraries, unity as *difference* and not as *identity*. Thus the theoretical a priori identity of contraries will itself contain (in the mode specific to it) the a priori yet "nonthetic" moments, corresponding to equivalent moments of the philosophical decision. This decision includes four fundamental moments: 1/ the *manifold* or multiplicity of terms; 2/ the *transcendence* or exteriority of one term in relation to the other; 3/ the *position* or positionality of each term; 4/ the superior *unity* of the whole that each term assumes for the other. To these four moments will correspond four nonthetic and thus non-philosophical a prioris; these a prioris are characteristic of the theoretical and define in some sense the a priori *scientific representation*. Theoretical description proper or the production of the object of knowledge will thus pass through four stages. It will consist in reformulating or redescribing the philosophical dyad—which is now a material—as an immediate or non-philosophical identity of contraries. This identity, far from being simple and having a single mode, operates according to the four previously distinguished modes. These modes have each time the nonautopositional and nondecisional form of *Identity-of-the-last-instance*.

Non-philosophy as first science thus allows us to describe science's most general rules. They can be applied to every term or statement, provided it is grasped *with* its virtual horizon of philosophical operations and decisions. We need not, however, carry out all the de jure descriptions and be exhaustive, because they implicate one another, and it is enough that *one* of this type, a non-philosophical description, exists for all the others to exist virtually. We are in science: to produce a validated knowledge is enough, in a sense, even if its constitution or its rectification is necessary.

The produced statements no longer describe the World and, above all, nothing philosophically comprehensible or organizable: abstract statements, unimaginable for perception and the philosophy that extends it, which have a meaning, but not the philosophical or rational meaning. They have a universality or a theoretical content superior to philosophy's, and the different philosophies can be exchanged within them as in the dictionary of a universal translation. In effect, the Identity-of-the-last-instance of terms X and Y breaks their bi-univocal correspondence, their parallelism, and makes visible the necessary change in postulates. More than that: the change in the type or nature of axioms, since the second philosophical postulate (the specular dyad of terms X and Y) determines the nature of the philosophical "axioms" (the dyad is *determinant* in philosophy and not only *dominant*). The replacement of philosophy's anhypothetical hypotheses with true scientific (yet real) axioms liquidates every parallelism or doublet in thought; it is the equivalent of a non-Euclidean mutation.

Given its relation to philosophy, its radical transcendental nature, the science of the One discovers as scientific a new domain of realities. It gives a theoretical status of *data* or of objective givens to the phenomena that fill this region and that did not have such a status before it: "phenomena" constituted by philosophy itself and its "natural" language. This science is probably the one that accesses most directly—we are not speaking of literary theory, which is not a science—nonlogical, nonformalized language. It accesses this language from the perspective of its meaning and not of its signifying materiality, which is the object of linguistics. (But this linguistically formulated materiality can be regrasped in turn through its virtual philosophical formulation and can enter into the sphere of the new science.) For the first time, outside philosophy, which is not a science, a rigorous discipline of the most autoreferent language or discourse, the philosophical, becomes possible.

PART II

THEORY OF GENERALIZED FRACTALITY

3

OF DETERMINATION-IN-THE-LAST-INSTANCE AS DESTRUCTION OF THE PRINCIPLE OF SUFFICIENT DETERMINATION

THE PRINCIPLE OF SUFFICIENT DETERMINATION AS HORIZON OF THE PROBLEM OF SINGULARITIES

LET'S return to the problem invoked in the introduction: the *reality content* of knowing, of science, and of philosophy. Its solution will introduce us more concretely to the problem of fractality.

We can call philosophy, and the specific history that accompanies it, the series of watchwords, injunctions, or balance sheet programs that take the following general form: return to the real! return to things themselves! at last thinking the forgotten real! and so forth. The balance sheet or digraphic form is decisive here for identifying the philosophical style: previous philosophers lacked the concrete, *this* particular determination and thus *the* determination eluded them; the fold or catastrophe that is in the real and that forms a singularity remained hidden from them; the elucidation of this lack, of this failure is the authentic real and the task to come, and so on. The philosophical decision is a retrospection of underdetermination and an anticipation of overdetermination; it is a decision in view of sufficient determination. There is no philosophy that does not echo this complaint and is not animated by this hope.

We suspect the following, and this is all we can say provided we stop philosophizing naively: the philosophical "real" was only ever the synthesis of this deception of an ostensible real that escapes and of this will of an ostensible real that looms, the *decision* of the sufficiency of determination. This decision is itself doomed to an insufficiency and an indetermination of a new kind. It is the whole of philosophy that "lacks"—but this time in an absolute and unconditional sense—the real because it desires it and because it believes it did enough for the real by desiring it. Philosophy begins by dividing and thus losing the real; it subsumes the real under operations that are at once too large and too small for it, too general and too particular. Philosophy as such is active ignorance, the repression of the following law: the real is not a contrary that flees before its contrary, like coldness before warmth; rather, it flees before the very pair of contraries and before every decision. And it is merely a subtler form of this ignorance to conclude, like the Contemporaries, that it is the impossible or the undecidable. For the real does not even flee before the pair of contraries. It purely and simply ignores them, and this defines its philosophical misfiring in a very different way. Hence the futility of the interminable labors that consist in attempting to concretize philosophy by impregnating it with aporia, contradiction, structure, difference, *différance*, the undecidable, the fold, catastrophe, language, or else logic—we can recognize, in this hunt for determination, philosophy's contemporary history and its vain attempts to take hold of the real or else to let itself be put to the test by it. The love of singularity is not singularity, but remains the love faced with the indifference of singularity. Philosophy itself exists only insofar as it abolishes the determined and replaces it with the will-to-determination. The watchwords that punctuate the history of "Greek" and "Occidental" thinkers were not really destined to make visible within philosophy an activity that develops outside-the-real, i.e., completely and without exception in the veils of transcendental Appearance, but to make us enter better into philosophy and under the law of its appearance: the call to abandon a philosophy unfaithful to the concrete is merely the decoy deployed by another, equally unfaithful philosophy in order to recall the dissidents under its banner and thus under the philosophical Authorities in general.

It is in the very essence of the philosophical Decision, in its *concept* of the real, that we must search for the reason for its radical inability to think the Determined or singularity. We can in this way gain a better understanding of the two great historical measures it took, in the epoch of Modern Times, to justify its approach as the only possible procedure: 1/ Critique of the *determined* as dogmatic and of the dogmatic concept of the determined—new primacy of *determination* over the determined (Kant); 2/ Autoposition of determination, which is thus presumed to apply to the real; hence the affirmation of determination's sufficiency (sufficiency to the real): the *Principle of Sufficient Determination*. These two traits are possible in their turn only by virtue of philosophy's oldest operation, *the sameness of the real and of thinking-as-logos, then as-reason*. Hence the "Principle of Sufficient Reason." In reality, reason can be presumed to be determinant and sufficient only if it is thus inversely *associated* with a real function or if it receives a "transcendental" value; only if the logico-metaphysical is conflated with an instance of reality. Hence the amphibology of the logical and the possible with the real, the mixture of the metaphysical with the transcendental: this is the ultimate foundation of the Principle of Sufficient Reason. This latter is necessarily filled by the principle of *sufficient determination*, which is deployed nowadays in an increasingly catastrophist pathos (turn, fold, catastrophe, etc.), because determination—a rational operation—becomes sufficient (to the real) as reason itself (reason is here regrasped in its phenomenal plenitude, whatever its mode may be—analytical, synthetic, dialectical positive, differential, etc.).

However, the real is no less repressed in this case, because Determination annuls or suspends the Determined. Its transformation into a principle ends up rationalizing it, dividing it, and dissolving its reality. It is thus not the "meditation" of the Principle of Sufficient Determination that guarantees its irrepressible insufficiency . . . Nor its deconstruction, since a supplement of "becoming-other" cannot by definition make this principle globally contingent. Deconstruction's half-solution makes clear that it places itself under the authority of the Principle of Sufficient Determination. The real critique of the philosophical enterprise can be carried out neither through a *supplementary decision-of-return to the real* nor through *the undecidable (real) that sends us back or "sends" us* (Turn-without-return) to the little piece of the real left in representation. These two attempts continue to be repressive and to seek to inscribe the Determined in a Determination too wide and too narrow for it. They seek to divide and exhibit it; and they refuse to recognize transcendental Appearance's scope and depth.

The only solution that does not enter into the repression of the real and that allows us to "analyze" it consists in grounding the radical contingency of the philosophical decision and of the philosophical undecidable, i.e., of

Determination, in the Determined itself. On one absolute condition (its historical importance will become clear): that the return to the primacy of the Determined over Determination not be simply a "return"—one more return—to dogmatism against the critical-transcendental style, but the beginning in a given or in a Determined sufficiently autonomous that it is no longer understood as a mode of reason—of sufficient reason—and that with it the transcendental dimension of thought is not lost, that this dimension is, on the contrary, included in it. We have to discover, against dogmatism itself, the Determined as a transcendental immanent experience, sufficiently determined *in* itself, and to take it as a guiding thread or paradigm of a new practice of Determination itself. Instead of going from Determination to the Determined, we will pass irreversibly from the latter to the former. Correlatively, Determination will no longer be a task before us, a destination or a teleology. It will now be the absolute aprioric opening without internal delimitation, an opening that is deduced from the Determined and expresses its effect without conditioning it in return. It is the ensemble of the a prioris or of procedures devoted to subjecting the Principle of Sufficient Determination to the law of the Determined: *to determine the rational determination by the real.*

When it is not simply the Determined itself, or else determination *as it flows now from the Determined*, "sufficient determination" is a mode of the founding amphibology of the philosophical decision, an amphibology of the real and of reason. To this confusion, we oppose the solitude of the real as Determined and its power to determine—in-the-last-instance—reason itself. It is not reason that determines the real; it is the real that determines reason. It is not the real that is the Other of reason; it is reason that is the Other of the real. It is excessive and transcendent—this is the genuine dogmatism, the most profound, of every philosophy—to say of the real that it is rational. The most we can say about the real is that it is the Determined *and therefore* the determinant and that in this capacity reason obtains its determined-being from the real and from the real alone. Provided, as we have already noted apropos dogmatism, that the "retrocession" of determination to the real at the expense of reason not be the simple inversion of the philosophical order, but the consequence of the fact that the real is already in itself absolutely determined—in the way in which the Last-Instance is determined.

If the philosophical decision is a hastily determinant and thus inde-
terminant experience, since the watchword of sufficient determination is
what divides and loses the Determined, the new paradigm can no longer
correspond to any watchword whatsoever or to an injunction of the type
"take care of being as a whole" or the type "return to things themselves!"
It is the "contemplative" way of thinking, which "starts" from the cause
and *occasionally* from things in order to go toward representation and
its labor without the shadow of a hope (and of a necessity) of "return."
This thought discovers in itself the power to suspend philosophy's under/
overdeterminant decisions and puts an end to the authority of the mixtures that associate any determined whatever with a decision. Philosophy
believes it can discover the Determined through mixtures at the end of
determination, as its concretion or its global effect. The deconstruction of
the Principle of Sufficient Determination, like its "postmodern" autodissolution, belongs to the philosophical exploitation of the Determined and
to the contemporary implosion of the transcendental illusion. It has no
validity for describing chaos, and even for thinking phenomena as "materials," *irreducibly* singular, dispersed, deprived of ideological envelopes.
These phenomena are perhaps none other, as we will observe, than scientific knowledges.

We hope to rediscover in this way reason's real phenomenal content: not
such that it would be codetermined by reason, but what should be called
its *real basis*, an instance that has the power to transform and generalize
reason into positive "nonreason," a real content concealed by the "principles" of reason (the principle of sufficient determination, of maximum and
minimum, of indiscernibles, of the "I think," etc.), which express its vicious
autoposition.

THE DETERMINED'S PRECESSION ON DETERMINATION

The Determined must be given and "usable" as such before every operation of determination and thus of donation. An operation cannot itself
provide the real and singularity: it has to obtain them from elsewhere,
i.e., from themselves. From itself, it introduces at most division, i.e., an

under-determination and a poor universalization, a loss in singularity that does not do justice to the Determined. The Determined, on the other hand, is included "in" itself, lived not "in itself" but "in-One," that is to say, as real Identity. Unlike the logico-real Identity that philosophers construct on the basis of the Principle of Identity, *real Identity is deprived of every operation, of every transcendence; it is the One, which we have described as nondecisional as much as nonpositional (of) self. The ground of the real is of the order of the strictly immanent lived experience and, what amounts to the same, of the order of the radically "indivi-duel" that we call "indivi-dual."* The Determined is already given, given by itself rather than acquired. It cannot be transmitted or produced; it cannot be created from scratch by an operation of external distinction and assemblage. It contains no determinable matter within it. Just as the individual—provided it is understood in its "non-philosophical" sense—precedes every process of individuation that cannot affect it, so the Determined—*as nothing-but-immanence or provided it is understood in the sense of objectively transcendental, nondogmatic lived experience*—precedes every process of determination exerted on the Determinable and cannot be produced by such a process. The Determined is the Undivided itself, the real Identity, which has never been acquired and given in exteriority, even as *causa sui* or in the wake of an infinite process. What is unlimited, on the other hand, is Determination's work on the material of the Determinable.

From the "first" Determined flows then a new concept of Determination. The suspension of the amphibology of the real and the rational—or rather the suspension of its claim, since the mixture is conserved in its effectivity—cannot be once again an external operation, a simple decision or separation, a *unitary duality*. It has to be—between the real itself and the amphibology of the real and the rational—an absolute, originary duality that is not produced on a prior identity nor destined to be unified or closed. A primitive duality, an indifference in a unique and non-reciprocal direction. Or a double donation of the real: as the nonblended, as the already-Determined in itself, *and* as blended with the rational in the style of philosophical mixtures. A duality constituted of the Determined and of another pole, that of Sufficient Reason, already reduced or suspended by the previous one, that is to say, in this case, rejected by it as unreal and transcendent. The "logic" that should be recognized here is no longer philosophical; it is the logic of the Duality-without-scission, as it is founded in the One: logic of the *a-priori-separated-being*, which flows from real Identity and holds philosophical mixtures in abeyance. Far from amphibology positing itself as the essence of the real, it is now only its most transcendent side, the most deprived of reality. The new paradigm makes the real-without-reason what determines—suspends in its claims but also realizes or determines—reason-without-the-real: the presumed determinant reason must now be determined and become real, and it can become so, given the starting conditions, only as *nonreason*. Nonreason is neither the negation of reason nor the reason of the no(n), but the correlate that corresponds to the real or to the One. It is reason when, ceasing to claim to determine the One, it is transformed into a simple correlate-of-representation (of) the One that determines it in this function.

Marx foresaw and masked the a priori, static duality of the sufficient real and of the de jure insufficient reason. The duality of the intra- and super-structure is *of the type* of these *real dualities* that serve to replace philosophy. But Marx assumed this duality without grounding it in the elucidation of the real's radical or antephilosophical essence. He demonstrated—against the whole philosophical and "idealist" tradition of the Principle of Sufficient Reason—the insufficiency of philosophical Reason in its "real content." But he was unable, strictly speaking, to found this demonstration on the precession of the Determined on Determination. He did not begin by elucidating the essence of the Determined and the nondogmatic sense of its precession and contented himself with conflating it once again with transcendent instances taken from the effectivity of History (labor power, relations of production, etc.). The first task of a thought of real determination—i.e., already determined before any operation—is thus to exhibit, in a specific and coherent mode with its object, the Determined *in* itself and to distinguish it from Determination and the Determinable (which is now "Reason," determinable as "nonreason"). Then, on this real basis, which is thus manifested in itself, it becomes possible to extract the a priori or *real dualities*, with their structure of inexchangeability, nonconvertibility, and irreversibility between the pole of Determination (with its rules that flow from the Determined) and the transcendent pole of the Determinable or reason. This is the labor of Determination become real, i.e., secondary or "without principle." We will describe it.

FROM RECIPROCAL DETERMINATION
TO DETERMINATION IN-THE-LAST-INSTANCE

If the Determined—elucidated in its radical essence—irreversibly precedes Determination, then Determination is a process whose sense and structure will change in turn. In the context of the Principle of Sufficient Determination, determination finds its completion and its concept as "reciprocal determination." In effect, when it is Reason that is (auto)determinant, it is so in a specific way that we should discern in order to oppose to it *the new practices of the already determined determination.* It is determinant

1/ through a *dividing and itself divided* causality (maximum-minimum, end-means, form-matter, agent-effect, matter-ideality);

2/ through a *reciprocal* causality of the determinant and the determined, the latter being determinant in its turn;

3/ through a causality that is *underdeterminant* (division) and *overdeterminant* (redoubling, doubling, and double); the completed determination—in this case the determined—is the result of a synthesis or an *effectivity;*

4/ through a *spatial* causality *of the topological type* at best (continuous or more or less dehiscent fold of the determinant and the determined; for example, fold of "cause" and "effect"), which then continually conditions all the derived causalities (sociological, economical, psychological), etc.

On the other hand, when the real or the Determined (nondogmatic, i.e., not rationally determined) is determinant, Determination becomes a new form of causality unknown to metaphysics. Marx divined this form under the rubric of "determination in the last instance." The Determined or immanent Identity is determinant:

1/ Through a causality that neither divides its object (here "sufficient Determination," which is driven back to the state of simple Determinable) nor, above all, is itself divided. Nondivided causality of the "last instance," which remains in itself, regardless of the distancing of what receives its effect. Unidirectional or irreversible causality, always exerting itself from the real toward the mixtures of effectivity, from the Determined toward the Determinable, without taking a path of return and without looping back.

The Determined does not share Determination with the Determinable, to which "overdetermination" alone will return: through a causality that acts each time on the "whole" as such of the Determinable. It treats this whole globally, with rules that respect a priori its indivisibility—its identity—and do not attempt to dissolve or deconstruct it afresh.

2/ Through a causality of extraction and not of abstraction. Extraction of a priori structures, which form the "tissue" of the real's scientific representations. In the ambiguous term *determination* we will distinguish a noetic side (precisely this causality of Identity, the "determination in-the-last-instance") and a noematic side (these a priori structures that "fill" or specify the previous side and serve in their turn to determine the Determinable).

3/ Through a causality that determines a function of "overdetermination" for the Determinable (without the counterpart of an underdetermination). It globally transforms the Determinable into a condition of existence (although not an essential condition) of the a priori structures of Identity's representation. It does not divide or underdetermine the Determinable, but uses it as a necessary condition for specifying and individuating the a priori, i.e., for "determining" it in the only mode in which it can be determined (the a priori that is already determined by its cause): the mode of overdetermination. We will be attentive to a few nuances:

- since the a priori is already determined, it can at most be overdetermined by the Determinable itself;
- the real is what realizes the Determination, and this Determination consists, among other things, in liberating the Overdetermination and "abandoning" it to the Determinable. To determine the Determinable is thus also to transform it into simple overdetermination. The system Determination-Overdetermination cannot produce effectivity, philosophy's weakened singularities, but something like hypereffectivity, knowledge's dual singularities. The effectivity or the Determinable is taken integrally to specify-individuate the a priori structures that will have been extracted from it through the force of the Determined.

4/ Through an absolutely nonspatial or nontopological causality. The exclusion of every (idealized) space outside thought is one of the most typical traits of the new paradigm. The rejection of *determinant distance*—which belongs to every rational or sufficient determination—in the sphere of the

simple Determinable, a rejection by the Determined itself, ends up liberating thought from its toplogico-transcendental model and its variants (the
figurative and the figural; the fold and the catastrophe, etc.). Understood
in this radical sense, the Determined is not only unfigurable; it is the very
force of the Unfigurable, that which transforms every figure into a "nonfig-
ure" (where the figure is now used for overdetermining its condition or its
unfigurable real kernel), every topology into a "nontopology" more universal than the topological determination of thought, every catastrophe into a
"noncatastrophist" experience of catastrophes, etc.

The materiality of phenomena, their singularity, has ceased to be
attached to a materialist thesis, to a philosophical position and a philosophical decision. It is now an ensemble of procedures—defined by rules that
formalize this operation of Determination-Overdetermination—exerted
on the Determinable, i.e., eventually on phenomena that demand a strong
coefficient of "empirical" materiality. We will no longer conflate the materiality elaborated, the singularity acquired or produced dually on the real
basis of the One, with the objectivated or empirical, always transcendent
and mixed-unitary forms of a brute or primary materiality. All the philosophical watchwords of return to the concrete or to things themselves—
whether it is a matter of the "perceived World" or "Being itself," "Reason
in history" or "Arche-writing," "postmodern" debris or "desiring flows" of
the unconscious—have something in common: they seek to place the most
fulgurating singularity of the real alongside, in the vicinity of, withdrawn
from . . . objectivation. Understood in its phenomenal richness, objectivation is a structure of thought whose invariants we have described, where
every phenomenon to be interpreted is presumed to be accompanied by
a space of interpretation or a universal plane that traverses it, separates it
from itself, and stretches it out for an infinite becoming. This paradigm of
thought is obviously the philosophical Decision, articulated and founded in
a postulate that elsewhere we called Heraclitean and that we can tentatively
compare with Euclid's fifth postulate ("parallel postulate"), with which it
has some affinity. In effect, when the Determined is elucidated in its mode
of radical donation, from which is excluded every quasi-topological plane or
space of interpretation that becomes indifferent to it, this plane—or rather
all the possible types of such planes become contingent in relation to the
Determined itself and move into the sphere of the Determinable.

One may obviously wonder whether we have not replaced the topological codetermination of thought with another, attaching to the thought (of) the Determined—but this time outside it—a "space-time" of a new type? Probably not. A crucial difference is that, on one hand, this "space" is at last radically abstract, free of any determinant reciprocal topology or distance, stripped of every transcendental geometry; and that, on the other, it no longer codetermines the Determined, but is "seen-in" the Determined, in its immanence rather than from itself or through a process of infinite torsion-reflection. This opening without "decision" of opening, this unfilled space of an "open" or of an ekstatic-horizontal position, this uprightness in a word is what we call uni-lateralization. Unilateralization is thus an operation of determination exerted on the Determinable, but uni-rectional and without possible reciprocity.

The new paradigm excludes, as we see, the whole manifold of catastrophes with which the philosophical decision is encumbered and that are modes of the invariant of objectivation. Unilateralization is precisely objectivation's phenomenal or real content; it does not have its form, but can, at most, as we now know, be overdetermined by it. Unilateralization is the a priori—itself complex and articulated—that is necessary for there to be "real objectivity," rather than "objective reality," and an authentic experience of the object, for example, scientific or aesthetic, that no longer responds to the mechanism of philosophical objectivation and topology. The "nonphilosophical" paradigm shows that the object exists, but without passing through the system of decisions and doublings by which philosophy *ensures* the object in its own mode and annuls the determination by seeking to realize it through doubling. Unilateralization represents something like a diachronic, irreversible objectivity, without chiasmus. It is the outcome of the "dualysis" of the decision's mixture, the real duality of nonthetic objectivity and of objectivation. *Objectivity* is an old word that harbors an amphibology and conceals unilateralization's phenomenal existence. Unilateralization operates in the manner of a "reduction," but without recovery or redoubling. It is finished and static like duality itself—it has only its process—and has already reduced every philosophy to the state of determinable materials. It is indeed an experience (of) transcendence, but a "simplified" experience, one that ceases to be divided and folded on itself. It is the only form of transcendence that can be correlated to immanence when this immanence is thought as radical.

What we can call Determination in a new sense of the term is thus this unilateralization, with its "complement" of overdetermination. Unilateralization defines the real aspect of singularity; overdetermination, its other aspect, constituted by the Determinable that is assigned to the functions of Overdetermination. We thus abandon the old transcendent paradigm, which projects the singularity at worst on the object's generality or screen, at best on the "back-fulguration" that illuminates the screen and that, instead of extricating or removing the singularity from its transcendent coating, has not ceased enveloping it more and more.

THE TWO DIMENSIONS OF UNILATERAL SINGULARITY:
1. THE DETERMINABLE, CHÔRA, OR UNIVERSAL MATERIALITY

Unlike philosophical singularity or singularity-by-mixture, the "non-philosophical" singularity is essentially unilateral and rests on the Determination-in-the-last-instance. We still have to develop its two sides in their specificity—both of which are (even the Determinable) an internal requirement—into a prioris of science. For the Determinable not only designates the philosophical Decision's brute given, but also what it becomes qua givens or *data* of science: it is an a priori and is equally determinant by its a priori status.

In the philosophical regime, any object (sensible or ideal) contains two entangled sides, both dissimilar and partially identical. Their identity is equivalent to their interinhibition or their mutual restriction (for example, to the visible side corresponds an invisible side that delimits it; to the given, a procedure of sedimented and enveloped production; to the external horizon, an internal horizon; to the signified, one more and one less signifier; to the full case, an empty case, etc.). This is an essential law; it can be indefinitely exemplified. Even the Heideggerian "withdrawal," which is "the same" as donation, remains an (extreme) mode of this law.

On the other hand, we move to the unilateral singularity, at least to its material or "determinable" side, by absolutely unfolding—without supplemental refolding, operation of folding or fold of Being—the two sides of the mixture and by inserting them within a "space" of another type, absolutely simple or foldless—more powerful or more universal than these sides and that manifests them in the same way. Not only is the empirical or objectivity "inscribed" in this space, ceasing to be simply something empirical, but also the withdrawn or invisible side, which concealed for philosophy the set of strata, procedures, horizons, and processes of objectivation, of decision and position, the set of conditions of possibility and even the "withdrawal" or the "reserve" that are also given or manifested in this space. They are now manifested in an absolute way (we suspect that it no longer corresponds to a philosophical "parousia"): they will be seen-in-One-in-the-last-instance. The a priori's transcendence and ideality, and even the transcendental withdrawal, whatever its mode, belong from now on to what we call a *universal or nonthetic materiality*. Not a restrained or mixed philosophical materiality (opposed or contrasted to ideality), but a materiality that is formed of the absolute manifold of all that can be given in a sensible and ideal way at the end of transcendence: all that is not given in the mode of the One's radical immanence. But it is also not an "empircization" of ideality or a horizontal "flattening": it is the insertion of the horizontal-ekstatic itself into a material universal "expanse" devoid of any structure of objectivation. It is thus the *object*, if you want, but more powerful and universal than the procedures of objectivation, since they are in turn reduced to the state of this specific "object" of science. In other words: the absolute, reserve-less donation affects the reserve itself; it is a materiality without a materialist thesis to limit it, since the materialist thesis is now included in this materiality.

Two contemporary interpretations should be carefully avoided. This universal materiality cannot be *textual* in its order: the textual (even the "general" textual) is precisely a mixed mode of singularity, in which the material side is hastily filled by transcendent phenomena (signifier, arche-writing) drawn from the philosophical object or its margins. The textual generality is restrained or limited by the signifier's empirico-material side (and vice versa) and must be reversed in its turn in a more universal and more abstract materiality. This materiality will include it as a simple material or determinable, in the same capacity, without any privilege, as the empirico-material side. The dual singularity introduces a more radical experience of the determinable and emancipates thought from textual servitude. But it equally emancipates it from another conception: *surfaces*. Nonthetic materiality is not obtained through a process of *rise-of-surfaces* or through a universal planification. To be sure, surfaces and planes are said in the plural, and even "in the multiple," but they interbelong, conceal one another or become-screen, mutually "save" or "spare" one another. These procedures of difference and identity pertain to philosophy or to mixtures and have to be transferred in their turn to universal materiality; they have to be treated as a simple material and manifested outside every *surface-of-manifestation*. Contemporaries have developed the style of "surfaces" and forces, of desiring surfaces or textual surfaces, of desiring forces or textual forces. But the surfaces are generalized only through the most invariant philosophical operations: dividing/doubling; sundering/refolding, etc. To these mixed, semimaterial, and semi-ideal surfaces, we will oppose a materiality capable of encompassing ideality itself, a materiality that is not further restrained or limited by its empirico-objective contents. We will not reach a theory of truly universal singularities unless we succeed in breaking their *limitation* or their *specification* as textual, topological, desiring, semiological . . . their nature as mixtures.

This is to say, that within the material space everything is indifferently yet positively reduced to the state of the determinable: all the philosophical differences and hierarchies; all the degrees of reality, of sense and of value; all the articulations of the given and of procedures; all the syntaxes of philosophical decisions and their specified modes (for example, by Human Sciences) are de-posited as indifferent, determinable and, to this extent, equivalent. Everything is now simple "materials" for other rules: those of Determination-in-the-last-instance that stem from the Determined or the One. With regard to the Determined and Determination, the Determinable is composed of the two sides of the mixture outside their transcendent relation, sides that have become indifferent to themselves and thus equivalent. Even the philosophical decision becomes a philosophically inert material. Such a materiality can no longer be limited or reduced to its empirico-objective form because this materiality is only a "point" or an indifferent manifold without privileges. It is simple rather than simplified, without a structure of decision and of (re)doubling. It is an absolutely "non-baroque" singularity.

Thus science does not posit the object *twice*, because it includes the auto-position in a new, more universal experience of the object. But what does it mean that science "requires" this? It is the Determined itself (on account of its precession on Determination) that "extracts" its two sides outside the mixture. It renders them indifferent and equivalent and casts them into a new place. In a complementary way, this extraction has another side: the suspension or reduction of the mixed form, its pertinence, its claim to legislate, and its insertion in its turn into the Determinable. In reality, this double "operation" is not an operation; it is simply what is seen and described of the mixture in the immanence of the One. It amounts to seeing the "extracted" itself rather than proceeding to an *effective* extraction and, on the other hand, to seeing-in-One the mixed form as suspended or reduced, rejected as transcendent and unreal. This nonoperational character of extraction and of reduction forms a system with the precession of the Determined on Determination and of the residue on reduction.

We can call *chôra* this space of universal materiality. A space that is never devoid of matter, but truly universal and devoid of topological structure, of fold and refold. Grasped in this way, outside its geometrico-philosophical determinations, and having become as *a priori of materiality* the object of a transcendental experience, the *chôra* is clearly the *equivalent* of the Kantian "sensible intuition": it *gives the object* or the phenomenon. But it gives this object in an absolute way, since it equally gives in this mode and indifferently the a priori forms of philosophical or transcendent objectivity that Kant, for instance, distinguished from their matter. There will be no real *universalization of transcendental Aesthetics* unless thought accesses an experience of the object's materiality that is not limited by the apparatus of objectivation, unless it stops wanting to fill intuition's pure space and time with local geometric knowledges. Dogmatism's end will be assured in the same stroke, the end of sensibility's *unitary reduction* to the understanding, thanks to the *dual reduction* of the mixture of sensibility/understanding for the benefit of a materiality that will not be limited in the sensible-empirical way. The scientific posture is neither Leibnizian nor Kantian. It reduces every philosophical decision, i.e., every variously proportioned mixture of sensibility and the understanding, and offers a universal Determinable to procedures of Determination—a Determinable that is the object of a transcendental experience.

In the *chôra*, which is itself "extracted" and "reduced," all the transcendent determinations become equivalent. This material space free of every geometry, which reduces in a certain way every scientific knowledge or every theory to the state of a radical empirical [*empirie*], is equivalent to the *emplacement* or the "em-place" of these knowledges as well as of philosophical decisions that take hold of them. The Em-place is not the philosophical *displacement*, which is always coupled with an anterior *reversal*. It is the absolutely originary and simple place, the topos-without-topology that the scientific posture requires and offers to every philosophical decision. The Emplace is the Determinable's absolute manifestation as such. Within it, all the determinations become of the order of indifferent material.

There is thus an absolute materiality or transcendence, a nonempirical manifold or multiple, identical in-the-last-instance and therefore not accompanied by a doubling multiplicity. The *chôra* is a materiality that is radically individuated as materiality. It is the manifold of transcendence, but it is an extracted and reduced manifold, a transcendence that is transcendentally extracted and freed from its philosophical or mixed forms, the forms of autoposition. It is not a site outside generic and specific distinctions, but *the* site, the topical manifold of and for these distinctions.

THE TWO DIMENSIONS OF UNILATERAL SINGULARITY: 2. THE DETERMINANT OR UNIVERSAL OBJECTIVITY

From the mixed to the dual or unilateral, singularity's material side gains a nonpositional universality. The same is true for the object's ideal or objective side: there is a "real objectivity" rather than the mixture of "objective reality"; it too must be developed for itself, unleashed from its premature blending with the object and rooted in the Determined alone. If we take the Determined as an immanent guiding thread, then a pure objectivity (without any object or any blending with the Determined) will manifest itself in the state of absolute given. Just as the Determinable is extracted from its mixed form, and just as this form—its authority at least—is reduced and rejected as unreal, so this "Determinant" is seen-in-One or extracted from its objective-mixed form, and this form is reduced or suspended as impertinent (at the same time that it passes into the "simple" universal materiality). The residue thus extracted is clearly a priori as well—like the Determinable—and it is the object in-the-last-instance of a radical transcendental experience.

Concretely, it is a matter of the abstract and absolutely universal form of objectivity, at least of its phenomenal residue. Given the Determined's precession, this form is by definition devoid of the structures of autoposition and decision, of the dyadic syntaxes of ekstasis or transcendence, of the project or the horizon. Absolutely open ideal milieu, purified of the traces of generic and specific distinctions, as well as the traces of every more originary philosophical operation. This ideality without object-rupture or internal limitation, without folding, doubling, or flow, is the objectivity = X (of) every object, nondecisional and nonpositional objectivity (of) self because it has the One as its essence. The *chôra* was the manifestation of the hidden itself, the absolute donation of the withdrawal in a mode that was no longer the opposite of the hidden ("parousia"); it was in some sense the philosophical object's "full-employment." To the same extent, scientific objectivity is in its turn the "full-employment" whose philosophical form is the restricted use. Real and not effective objectivity, but unreal in its nature, an unreal that has as such an absolute positivity in its order. This formal a priori of objectivity is now anterior to the mixture of the "objective reality"; it is extracted "by" the One and reduced in its effective form, voided in its turn of every philosophical topology.

Experienced and described in this way, the Determinant is simple only by its essence-One. In its specific nature, it is complex and includes three a prioris that correspond, in this nondecisional mode (of) self, to the a priori structures of objectivity: the object's transcendence or exteriority, its positionality or its stability, its identity or its unity. There are thus *three nonphilosophical a prioris of real objectivity:* 1/ a *nonthetic Transcendence* (NTT), experience (of) a simple, not redivided and redoubled alterity, without autoaffection; 2/ a *nonthetic Position* (NTP); and 3/ a *nonthetic Unity* (NTU), of which we can say the same thing as of NTT. This objectivity is in itself, by its essence, undifferentiated, in the sense that it applies to the most universal Determinable or constitutes a "space" that is absolutely universal, an ideal and no longer material space. But this a priori field of nonthetic objectivation is intrinsically constituted of different moments. It ceases to be a philosophically de-cisive, discriminating or differentiating factor, a factor of hierarchy, of sense and of value. It is "defatted" of every decision and forms a field of presence without a present object, an objectivity without ob-jectivation. To the *chôra* as intensification or densification of materiality thus corresponds an other side (which is autonomous in its order), the side of an ideality that is itself intensified by the suspension of all that is presented, not only as object, but as mixture of "objective reality."

Obviously Determination's complete concept contains not only the three procedures of nondecisional objectivity (of) self but also the *chôra's* procedure, which has the same nature. Given that these procedures are all founded in-the-last-instance in the One or the real's immanence, they are unified by this immanence with only the occasional help of an external or transcendent agent of synthesis. Together, they give place to Determination in the broad sense. Determination, as we see, includes not only Determination by nonthetic objectivity but necessarily this determination's overdetermination. It is ensured this time by the material or the Determinable, which represents the conditions of existence of scientific objectivity.

The edifice of the unilateral singularity is complex in its own way. It is no longer the bilateral or circular mixture in which determination is at once under-and-over-determination; it is the dual or, better yet, the unilateralization, through the determination (in a nonthetic form) of overdetermination (by the Determinable). Such a concept of singularity eliminates, once and for all, the philosophical effect of underdetermination by which the Determined's reality was dissolved a priori. When the Determined precedes Determination, Determination becomes autonomous and full in its order, not affected by a lack. And the overdetermination befalls it additionally and necessarily, but in no way as the wretched complement to an underdetermination. This is *complexity*'s positive concept, which philosophy does not know for the same reasons that it has no full and positive experience of singularity.

4

THE CONCEPTS OF GENERALIZED
FRACTALITY AND CHAOS

THE SCIENTIFIC GENERALIZATION OF FRACTALITY.
THE NON-EUCLIDEAN IDEA AND ITS THEORETICAL SENSE

A general method for the description of phenomena grounded in a theory of Identities is an idea that has to be philosophically limited and scientifically generalized. Differences, Catastrophes, Games, Multiplicities, Disseminations . . . also form such a method, but these concepts express the generality in an ultimately restricted and transcendent philosophical mode. They are indeterminate generalities that manifest themselves in half-singularities, and these half-singularities testify only to philosophy's ability to survive and not to its "adaptation" to the description of real phenomena. By contrast, we are seeking from the outset a form of generalized fractality (= GF) or chaos that is a *real* limit or critique of every possible philosophy; singularities that can be described in the spirit of science rather than as remainders of ontology's autodissolution. Before the "application" of the theory of Identities to philosophy, and after the expulsion of sciences' epistemological image, the intermediary task is to draw the last meaning of this new experience of Identities: to pass from Identities to the form of *real fractality* they make possible, to establish a *scientific thought of fractals* capable of generalizing their concept. Just as we previously described the adequation (of-the-last-instance) of science's essence and of nondecisional Identity (of) self, so we will have to describe on this basis the adequation of the Identities we will call *fractal* and of scientific representation, of the scientific instance of knowing. At the same time that we will reformulate a non-philosophical and nongeometric concept of fractals, which is generalized under the conditions of Identity and of thought-science, we will manage to liberate sciences from their philosophical image.

This theory of Identity, then of fractals and of chaos, can be generalized beyond every philosophical position only if it is drawn from the essence of science, from the scientific *posture*, rather than from the geometric "Mandelbrotian" concept of fractals. These fractals will serve us as materials, as the indication of problems to be resolved and as properties to be generalized; but they will not be the principle of generalization. This generalization is the Theory of nondecisional Identities (of) self. A direct generalization of geometrico-physical fractals, direct but external to their scientific sense, would produce a "philosophy of fractals," a "fractal vision of the World," which would be added to the philosophies of Multiplicities, Differences, and Catastrophes. How can we elaborate a truly universal concept of fractals if not by eliminating the geometrico-philosophical mixture, which would transcendently combine their limited geometric experience with a philosophical generalization of the "metaphor" of fractals? The authentic generalization of scientific givens must remain internal to science. It must not take the form of a hermeneutic and circular appropriation through a transcendent Tradition or a Reserve of thought. But it must also cease to transpose, metaphorically and continually, the geometric and physical givens of fractals into the philosophical space. As soon as its principle is absolutely internal (the Identity-of-the-last-instance) and as soon as it is exposed in the framework of "science itself" and not under the philosophical horizon, fractality *is* absolutely universal, in principle and as much as possible (more universal than its philosophical generalization-totalization). This does not mean that it is not subject to corrections. Quite the contrary.

How is this new generalization of already elaborated regional or "ontic" knowledges possible?

1/ Here the method of description of phenomena is general only insofar as it considers them from the perspective of their *Identity-of-the-last-instance* (identity-without-identification), rather than from the perspective of their philosophical *identification* to a transcendent instance (totality, mixture, dyad, doublet, etc.) as the Foundation, Being, the Concept, Differe(a)nce, the inconsistent Multiple . . . would be. It is a matter of relating them to what is less their "in itself" than a position of thought, the only one, *that renders intelligible the fact that their meaning is scientific, that they are products of science rather than objects of philosophy,* and that describes them in terms of its own internal requirements and not in terms of alleged properties, "objective in themselves." Phenomena's only real "in itself" is the cause of science and, in a sense, the very science of these phenomena. The Theory of Identities does not replace existing sciences as philosophy does (*if only partially*). Quite the contrary, it adds itself to them as the special science, which makes clear that the described phenomena have meaning only through and for science. This theory's "objective" is not to discover Identity, and the fractality that derives from it, as just one property alongside others (the "ontic" properties). It is indeed a question of a new property, but since it is neither ontic (the object of already existing sciences, like the geometry of fractals, for example) nor ontological and philosophical, it is a "unary" or specific property of "first science." In other words: *it is the specifically "scientific" property of phenomena; their capacity in-the-last-instance to fall under a science and what protects them from falling absolutely under the philosophical Authorities.* This special property—absolutely supernumerary and, as we will see, "fractal" vis-à-vis philosophy—is typical of the scientific representation of things and allows science to *think itself as the science of these phenomena.* Instead of an ontico-material or ontologico-philosophical property, Identity-of-the-last-instance (of) this or that phenomenon = x is what makes it possible to say that the alleged "properties" of X are, in-reality-of-the-last-instance, acquired scientific knowledges, produced and apparently reified in the form of *data* "in themselves."

2/ The second condition for a scientific generalization is apparently opposed to the first. But it is only its complement and, in some sense, its consequence. As to their mode of donation, the phenomena to be known must be considered as given at first under the reason of their philosophical sense or form, in any case as virtually philosophizable. The Theory of Identities (and thus the theory of fractals) can be a science of materials-being, of object-of-science-being of these phenomena, only if it is *also* practiced as a lifting of the resistance that philosophy spontaneously exerts against it. First science, in the form of Theory of Identities, is the real critique of philosophical singularities and multiplicities, which are forms of resistance against their scientific species and mimic science against its own spirit.

If it is possible to realize an equivalent to the theory of fractals in the nonmathematical sphere of pure thought, of "representation" in its universal form, a theory valid for natural language and philosophy and not only for some dynamic or physical geometric forms, this new discipline, far from being opposed in any way (less than philosophy in any case) to its mathematical form, will bring it instead a kind of confirmation—not "empirical," i.e., ontic, but transcendental or founded strictly on the internal requirements of scientific thought itself. But this amounts to saying that the theory does not depend in its existence on the mathematical and Mandelbrotian form. If the Mandelbrotian theory should by chance be invalidated, our theory of fractal would survive this improbable disaster, because the mathematical theory will have served as its material and indication and will not have directly served to constitute it in its essence (the generalized theory transposes fractality on a real, other-than-mathematical-or-physical basis, as we will elaborate it).

The Identity "of-the-last-instance" is the one that remains in its own immanence, regardless of the region of the real (the World, for example) in which it acts qua cause. Thanks to this radically "transcendental" property— to this essence—that is its own, it opens the possibility of a new type of ultraphilosophical generalization, which can be applied to philosophy but that has a scientific and no longer philosophical origin. Nevertheless, as science "of" philosophy (i.e., of the *data* or *phenomena* that philosophy is and uses as its materials in view of a science the One), it is not only that: it is *the science of the essence of sciences, of scientific representation*. It is not a simple discipline alongside others, but the one that thinks the essence of-the-last-instance of sciences and safeguards their sense as sciences from philosophical expropriation. This is why it represents, in two distinct modes, the equivalent of a non-Euclidean mutation in the experience of thought.

1/ This mutation does not take place at the interior of a science, but at first from philosophy (eventually a "philosophy of fractals") to a science. Clearly, this is possible only if every science is by its essence also a *thought* and can thus be compared on equal terms to a philosophy. The Theory of Identities does not represent an ontic or transcendent, but a radically transcendental mutation. Here we give a new sense to the obvious non-Euclidean "metaphor": nothing-but-transcendental and not empirically conditioned, scientific yet valid for first science or for the essence of every science.

2/ This mutation does not stay at the interior of an ("ontic") science. It equally takes place from an ontic science (fractal geometry) to a first science (neither ontic nor ontological, but unary). The generalized theory of fractals can be ultimately called non-Mandelbrotian, but in a slightly different sense from the previous sense of the *non*. It is not possible here to critique Mandelbrot's oeuvre, nor do we intend to do so; we do, however, intend to genuinely critique philosophy.

Since this mutation is not interior to science, since it does not go from a science to philosophy, but from an (ontic) science to another science (that of the essence of sciences), we will say that it is strictly transcendental, i.e., internal to the scientific practice, the only practice that is immanent and thereby distinguished from philosophy. It is less the fractals as supposedly "natural properties" that are generalized (extended, transferred to other regions of nature or to society, art, etc.) than their theory, which is also, strictly speaking, a *generalized theory* (of fractals). This is possible only if the essence of theory itself proves to be fractal in an other-than-geometric mode.

Because we are elaborating a radical transcendental concept of science (and thus non-philosophical, nonempirico-transcendental concept), we have to treat this procedure of non-Euclidean generalization in the same way. It undergoes a mutation that frees it from its geometric "Euclidean"/"non-Euclidean" division. In every respect, this new non-Euclidean use seems to be less metaphorical than philosophy's own use. It consists—if the analysis is pushed further—in liberating the *non-Euclidean Idea* from its restrained ontic forms and its philosophical metaphors and in restoring to it its *identity*, which can only be the identity of science, of the essence of science. On this condition, the non-Euclidean Idea becomes a universal theoretical tool.

It should be noted that the transcendental in question here is science's and not philosophy's transcendental. So it is equivalent to an instance of radical, undivided immanence, which is not split in a circle, doublet or empirico-transcendental mixture. The real cause of-the-last-instance is the only transcendental; and only Identity thus conceived restores to the transcendental its *identity*, which philosophy had lost.

It is science, science alone but as *first*, that is the internal cause and "beneficiary" of this operation of generalization. This benefit is not a capture, a slicing off, since everything takes place between sciences. Above all, we do not generalize Mandelbrot here the way philosophy has already done in order extract from this theory (which it did not itself produce) a surplus value of sense and of energy opposed to science. Vis-à-vis Newton, Boltzmann, Cantor, Einstein, or Mandelbrot . . . , philosophy proceeds

- by detaching and isolating a local theory cut off from its process—while we reinscribe it in the process of an (other) science;
- by its reposition as factum, autoposition, and rational transcendental fact that gives the sense or representation of Being—whereas fractal knowledges remain simple knowledges and are not identified by philosophical autoposition with the real itself, with "Being";
- by *philosophical* extension or generalization, i.e., by a divided (generality/totality) and thus limited universality—whereas the universalization of knowledges or theories is no longer affected here by a decision that would divide it, but rediscovers its identity—*of-the-last-instance*, of course. This is enough to open a theoretical "space" of absolutely unlimited rectification to the theory of fractals, a space that is not cleaved or closed by philosophical teleologies.

As Benoit Mandelbrot indicates with implacable theoretical rigor, the fractal is a "response to a question that does not even seem worth asking."[1] Obviously the question will be elaborated even better—this will be the fractal theory itself—if the response is already given. Nothing theoretically rigorous is carried out that does not start from a *given*. That fractality is *given* does not mean, however, that it has to be given in nature, according to the mode of existence of various irregular figures in the vital environment. If some knowledges were not also *given*, if fractality were not an emergent theoretical object "before" becoming a property of perceived objects, there would be no theory of fractals. The whole problem is to *adequately* identify and describe this precession and immanence of an object of knowledge, like GF in particular. The theoretical ordering of the problem (the theory can represent an object-of-the-last-instance without modifying it, provided it is determined by this object, which is not alienated in it) allows us to resolve this problem by abandoning the specular philosophical circle of the supposedly given fractal object and its supposedly given model. It allows us to escape the drawbacks associated with this conception: the idealist forcing of the object by the fractal model that abusively generalizes; the empiricist reduction of theory to an idealist abstraction of the object. GF's theory is not an idealizing extrapolation or extension of the supposedly given object (not a shred of philosophy is in any case given with this type of fractality). It is not an idealist attempt to intervene *in* philosophy and to *fractalize it under its own conditions*; GF is a relatively autonomous order of reality. Neither is philosophy a *region* of GF, nor is GF an intensification or else a deconstruction of philosophy (in both cases, its reaffirmation). There is an a priori fractal intuition; this is what must be described. We will not say that it is deduced from the One, but, more exactly, that it is the space of every possible theoretical deduction, starting from axioms that bear on Identity, axioms that operate with philosophical information, but that are no longer reduced to this information.

We hope that we have not produced a scientific ideology or a scientist's spontaneous philosophy or a philosopher's spontaneous science (as the philosophical or epistemological uses of sciences in general are). We hope that we have produced the *theory* of fractals, more radical or universal than the geometric fractals, but not "superior" to them in the sense in which philosophy presents itself as the "superior use" of existing things. We pass from ontic fractality to a unary fractality, whose concept is broader and applies to language itself (philosophy, poetry, literature, etc.) and no longer only to geometric phenomena. And thus it is not a question of a poor extension of the geometric to language, such as the one scientists sometimes practice, scientists who are seized by the most uncontrolled philosophical spontaneity.

SKETCH OF A DIMENSIONAL DESCRIPTION OF THE PHILOSOPHICAL DECISION: FIRST APPEARANCE OF A PHILOSOPHICAL FRACTALITY

Fractality is a problem of *dimension* and presupposes a prior dimensional description (adapted to these nongeometric disciplines) of science *itself* and of philosophy. Just as philosophy knows half-singularities and unfinished multiplicities, so it will know a half-fractality. This half-fractality will appear only if we can characterize the *dimension of the philosophical decision*. We will take this characterization of philosophy as our guide in the search for a more radical fractality adapted to the instances of science itself.

In general, *dimension* designates the minimum number of relatively autonomous coordinates or parameters that are necessary for representing an object (philosophy in this case). If the problem of philosophy's dimension can still receive an onto-philosophical sense, it can also already point toward another problem: *the specifically fractal dimension, the dimension of the most irregular manifold, a dimension that philosophy cannot thematize vis-à-vis itself.* What we are looking for is the *dimension-of-fractality* of philosophy itself. A dimension set in play by a generalized fractal relation to philosophy. The concept of dimension can be itself generalized and applied to philosophy through an adequate transformation, which is consistent with philosophy's theoretical possibility.

The dimensional description of philosophy was initiated in the affiliated form of *transcendental arithmetic*, which the Ancients in the Pythagorean and Platonist context employed to describe philosophy. This is the meaning of the great principles of oral and esoteric Platonism: the One and the Two, the Dyad—principles that can be easily transformed into the description of every philosophical decision. This decision is the most general invariant by which a philosophy *is recognized* and *assumed*. It contains first a *Dyad*, a coupling of contraries—precisely a "decision," in the narrow sense of the term (this decision gives its name to the whole, since it is decisive for it); then a *Unity* that reaffirms the coupling, that reposits *as such* the dyad-unity of contraries. This arithmetic is not mathematical, but transcendental; it claims to determine or constitute the real itself.

In this form, it suffers from a lack of true empiricism: it does not know, properly speaking, the 0 that serves to designate, in the dimensional description, the term or point—which it miscognizes, rejects as too empirical, or divides into a couple. It begins straightaway with a *couple* of terms, with the relation of the Dyad or with the 2. As a result, it suffers in a complementary way from an excess of philosophical or transcendent . . . empiricism, since what it calls the Two, by immediately tracing the existence of two points or two terms, the dimensional description would instead call the 1. In this case, the symbol of the dimension is not traced from the number of empirical or constitutive "dimensions" (what, for reasons of terminological clarity, we will call the moments, and not the dimensions, of the philosophical structure). As to the One, to the role it plays in the Decision, we will say that it is a third term or a second principle, or the "first principle"; but we will never-theless call it One and will use the arithmetical 1 to designate it.

A dimensional description of the philosophical Decision cannot confine itself to saying that the Decision comprises two terms + one. To be sure, it is already an important characterization to say that it contains two and/or three terms: the One is already included in a latent state in the Dyad and added to it as a supplement. Philosophy is, without doubt, a figure of thought whose dimension is comprised "between"—we will discover the sense of this "between" later on—1 and 2 (a line that tends to fill a surface, a half-line half-surface . . .). However, a more precise description would com-plicate the problem. For the Dyad of contraries does not have a dimension 1 as a line would, but a dimension 2 . . . In fact, the terms of the Dyad are *already* unified or intersected in a latent way by the One, which is not only transcendent to the Dyad but also immanent; external and internal to "expe-rience" (the Dyad). So much so the Dyad has instead the figure of a sur-face, a planar space with two coordinates starting from a point-One, which makes them relative to one another. The One itself, in its transcendence to the Dyad, adds a supplementary dimension. Philosophy has an intermedi-ary dimension between 2 and 3. Not an integer, but a fraction whose sense is obviously empirico-transcendental and not simply empirical or arithme-tic. We should compare this elementary description to the one Heidegger gives of onto-theology and its dimensions of *generality* (the plane of Being, ontology) and of *totality* (the vertical or the theological summit).

This ever supernumerary nature of philosophy, with respect to the cou-plings of terms with which it works, is significant and suffices in a sense to characterize its originality as an excessive, dehiscent thought, simultane-ously internal and external to the "givens" or to "experience." Nevertheless, an arithmetic that is simply projected onto a structure does not amount to a dimensional analysis. This analysis, even when it is applied to philosophy and becomes in some sense transcendental, will instead say the following: an isolated term has a dimension 0 (philosophy represses or does not know it); the relation between two terms or the Dyad has a dimension 1; the struc-ture Dyad + One has a dimension 2. *The philosophical Decision is an object with dimension 2, a surface or a plane rather than a volume.* But it is a matter of a mixture, of a surface or of a plane that is equally transcendental and not purely geometric. And no arithmetic or geometry exhausts its ontological content.

This surface- or plane-nature appears, strictly speaking, only via the dimensional description. It is however clear that this surface is not simple.

Already in its arithmetic characterization—let's start again from it for a moment—the philosophical Decision shows a fractionary nature in all its dimensions. Not only is the Dyad already fractioned (2/1 or excess of the 2 over the 1, its transcendence); but it globally corresponds to the fraction 3/2, which expresses the excess of the One over its own identity with one of the two terms of the Dyad—in both cases, the nature of philosophy's "transcendence." But in its dimensional characterization as well, its transcendental and not simply geometric nature means that philosophy is an "intermediary" being—not only between the dimensions 1 and 2, *but in each of its moments*, because the Dyad with a dimension 1 is an integer only in appearance and must already be represented as a fraction. In philosophy, the fractionary character does not lie in the relation between two structural moments but in the very nature of these moments.

This is to say that if it has the dimension 2, which is in fact a surface and is represented by an integer, philosophy is fractionary in relation to geometry, not only through the play, the gap between two consecutive moments, but intrinsically and in the very nature of its moments. It is so qualitatively and thoroughly. Not only in the sum of its moments, but in each of them where this sum is reflected: apparently, the philosophical surface or plane is intrinsically "fractal." The consequence is that philosophy already deals with a *certain* fractality, but is not an accidentally fractal object; that this special and no doubt limited fractality is not one of its properties, but its most internal essence. Within it the fractionary style is not "simply" arithmetic; it remains in the empirico-transcendental doublet, in the fraction formed from the transcendental term and its irreducibility to the empirical term, or in the dehiscence of the Other to Logos, etc., a qualitative alterity that distinguishes the moments among themselves, but also each moment from itself. *This is a sign: if we must seek a nongeometric (nonontic) and non-philosophical (nonontological) fractality, we can take as a "signpost" of its problem, if not of its solution, its philosophical concept and the dehiscence of the transcendental to the empirical or the "irregularity" of the Other to Logos, etc.*

So it is possible to conserve the concept of dimension for philosophy (and for science), but subject to its change in the direction of the transcendental. It then takes in principle a fractal or quasi-fractal value, and every dimension of the "philosophical" object is fractal as much as "topological." This explains a special property of philosophy: it contains at one and the same time more objects than the Dyad, which serves as its basis or its given, and no more, since each object of a philosophy can be placed in correspondence with a point of the Dyad. Philosophy is an intermediate being, a monster or a demon, at once identical and superior to the Cosmos. We can say that every philosophy contains no more things than the dyad of the Heavens and the Earth, but also that it contains at least one more: philosophy itself.

Another meaningful change that philosophy introduces into the geometric concept of fractals and that could serve as an indicator is another understanding of what an "intermediate" being is. In geometry, the intermediate between two dimensions (between the surface and the volume, for example) has no autonomous conceptual existence. The fraction is expressed quantitatively (by a number like 1.627, for example), and the reality or identity of this "fractal" property is referred to without being exhibited, if not in the form of another property, the internal homothety, which is another geometric knowledge. On the other hand, philosophy exhibits (no doubt in a still transcendent mode) *the identity or reality of this gap, of this fraction*, which is qualitative at least as much as quantitative: not only *between* 1 and 2, but *at once* 1 and 2. As an intermediary *being*, it possesses a *certain identity* and does not interrogate its origin, donation, and functions, as we saw earlier. It proceeds, nevertheless, by *identifying* the distinct dimensions, which end up converging at infinity: the objects with dimension 2 end up corresponding to those of dimension 1 and fill this dimension with their excess. There is a specific *reality* of the fraction as transcendental; and a first real reason, a first cause of fractality is provided in the form of a *transcendental identity* rather than a simple "local" property of homothety. This is how Nietzsche, for instance, seeing the acme of contemplation in the coincidence of being and becoming, creates a fractal object, but in the philosophical mode; or Heidegger with the In-between, the Fold of Being and being; or Deleuze with the Fold of the desiring machine-flow/partial object.

THE PHILOSOPHICAL SEMIFRACTALITY
AND THE CONDITIONS OF ITS SURPASSING

If the philosophical interiorization of fractality to the Decision itself as identity-of-the-Intermediate, as In-between or Difference, has the advantage of starting to exhibit the root of a transcendental fractality, it only surpasses ontic (geometric) fractality in the mode of its ontological appropriation, where Identity, instead of "founding" and guaranteeing the irreducibility of the fractal to the continuum, does its utmost to efface it. In philosophy it is Identity itself that is fractal, that is affected by the fraction and its division. This reversibility of intermediate-being and its identity produces a genuine circular disaster in which all parties are losers. Fractality itself undermines Identity, which is then no longer capable of sustaining it, falls back on it, and inserts it in a process of identification. Fractality injures the identity that exploits it. This game of diminishing returns does not lead to a nihilist, but to an overnhilist leveling, to a situation of equilibrium or hesitation: half-nihilism, half-counternihilism, a vacillation that is the way in which philosophy drowns fractalities in Differences, Disseminations, Multiplicities. . . . The real of fractality is no longer and was never Identity, but the circle of the Same or Differe(a)nce: restrained fractality, effaced as much as produced, "static" in the broad sense of the *Same*; tautological, if you like.

Philosophy will have merely realized a half-fractality, incapable of elaborating its most universal concept "once and for all." The most sure sign of its concept's "average" character is obviously the multiplicity of philosophical decisions themselves: each philosophy leaves aside, implicitly or not, a real it deems too irregular or particular to be rationalizable and masterable; a real that gives place to another, supplementary philosophy, which takes hold of it against the prior philosophy and is expressly constituted to rationalize it, but proceeds in such a way that its "decision" makes visible a new residue, and so on. Philosophical fractality, the fractality of each system, is limited by the concepts of fractality that other philosophies implicitly propose. The only unique and universal concept of fractality is therefore conflated with the hesitations and conflicts that compose the *continuum* of the philosophical Tradition.

Obviously the concept of an absolutely irreducible or non-negotiable "fractionary relation" should be elaborated. Fractionary: not only in the arithmetic-quantitative sense, or in arithmetic-transcendental sense of philosophy (one of whose avatars is clearly the "relation-without-relation" that Heidegger and Derrida speak of); rather, in the sense in which the "relation," the "fractionary" nature, which always contains an ultimate possibility of correspondence and continuity, would be itself ultimately abandoned. To detach fractality (at least by its reality or its identity) from the arithmetic fraction, and even from the philosophical "relation-without-relation," from *relation* in general, i.e., from the Dyad; to stop inscribing it in the form-dyad. To be more precise: "classical," geometric, or philosophical fractality is always said of figures whose irregularity derives from their *being*, i.e., from their *property* as intermediate natures, a property inscribed in the element of relations or relationships. These "in-betweens" were precisely coextensive with the Identities, syntheses, or connections that were transformed with them. So if we now seek a generalized fractality, we have to, if not eliminate every relation, at least no longer allow the "fractionary" relation and its identity—which are the two essential components of every fractality—to limit or impede *each other* and have to discover another "relation" between them than the relation of the Dyad (which is not only the initial given of philosophy, but its determinant essence). Those are the givens of the problem; what is its solution?

This program can be realized only if we manage, without any philosophical contradiction, to consider and treat as *given* (given beforehand and nothing more) Identity as an instance that is de jure absolutely irreducible to every "fraction" and even to every philosophical "relation." Identity is not or, more precisely, is no longer fractality. But it must constitute the condition of reality, the *cause* of fractality, which will maintain it outside its reductive inscription in a philosophical dyad or in a geometric "intermediary." As we have understood it, Identity is no longer even a *dimension* that can be connected to others. It is the *reality* of a nondimensional thought, but it is all the more fractal, because fractality is detached from its geometrico-arithmetic dimensional references which could only efface it. This identity no longer entertains any relation *with* (codetermination, reciprocity) something else, with its representation for example. It is the finally positive essence—experienced as such—of the "without-relation." Moreover, it will be able to afford fractality a new existence: a "relation" that no longer has the form of an identity and of a division—the division of this identity, precisely. Such a fractality will probably have no continuous relation to the philosophical Decision and its dimensions, to the half-fractality this decision is capable of.

A thought by dimensions (even by intrinsically fractionary dimensions) like philosophy gives place to a fuzzy, unstable fractality, deprived in any case of radical conceptual universality. The most irreducible and stable fractality is not intradimensional, but ultradimensional: *it is fractality "in" dimensionality itself, at least in its geometric-philosophical forms.* One/Dimension(s): this is the "gap" that cannot be reduced to any ever negotiable as much as non-negotiable gap. *Alterity*—if this concept can still be used here—no longer takes the form of a mediation, of an intermediary or of a mid-place [*mi-lieu*], the form of the universality of a milieu; it takes the form of a distance without return, of a unilaterality that is irreducible in proportion to its Identity-of-the-last-instance. To the fractality of philosophy, to the fractality of a circular dynamic system, we will oppose not a "linear," but a "noncircular" fractality—a fractality that is identical to a certain *order*, which cannot be reduced to what it puts in order.

The elaboration of a *fractal style* in the scientific thought of philosophy (a style rather than fractal objects) assumes that we have passed from a particular science, a regional theory, to a more universal discipline. But this discipline always remains a science and never becomes a philosophy. The only science that is more universal than regional (more precisely, ontic) sciences is the science of the One, which we call "first" for reasons already examined. This amounts to going from the plane of ontic objects or *data* to the plane of the "object" we term Identities. It is a very particular object because, giving itself in the mode of a radical or *in-the-final-instance* immanence, it enjoys two properties foreign to every object: 1. it remains what it is, without being transformed or alienated wherever it acts as cause (this is the main sense of cause or determination "in-the-final-instance"); 2. it is thus never given in the mode of the object, of presence, of representation, of Being that is transformed or alienated with its donation; it is not discernible in the horizon of philosophy's most general Greco-ontological presuppositions. In this sense generalized fractality, which will be a mode or a sequel of this Identity, will consecrate the break outlined in contemporary thought, not simply with "metaphysics," presence or representation, with "logocentrism" and its modes, but with philosophy itself. It is the *logic of continuity* and the half-singularities it tolerates that is discarded; the entire thematic of mixtures or blends, syntheses and co-belongings, of reversibility and of topological neighborhoods at best, of circular hierarchies at worst. It will be a question of a genuine *fractal opening* "beyond" the simply conveyed philosophical closures and teleologies.

FIRST SKETCH OF GENERALIZED
FRACTALITY'S (GF'S) CONDITIONS

Four invariant conditions have to be united for fractality "in general" to exist. They form a system in pairs:

- an irregularity, a fragmentation, an interruption, etc.;
- not arbitrary, but definable by a certain degree (to be fixed) of irreducibility to a continuity whose nature must itself be fixed;
- furthermore, a principle of constancy ("homothety") of this irregularity, a principle of self-similarity of fragmentation;
- which is itself defined in correlation with scale variations (magnification, degree of resolution, etc.).

These are indeterminate generalities, which serve as our guiding thread or as indications to be elaborated.

Generalized fractality (GF) unites those four conditions according to particular modalities, which distinguish it from its Mandelbrotian form. We will describe them progressively. Here is, first of all, a schematic chart of conditions that have to be fulfilled before we can move to a GF.

1. *The condition of constancy or identity*: to discover and identify, not so much the greatest possible inadequation between two terms, but the reason or cause of this inadequation, greater than any "gap," "differe(a)nce," "inconsistency," "dehiscence," "dissemination," "Other"; *thus of a nature distinct from every form of first alterity*. Since GF is "unequal" to the philosophical Decision, it must be unequal to all the forms of inequality, fragmentation, and partialization that philosophy can tolerate. In a sense, philosophy is virtually the most powerful logic. At least, it presents itself as the most autoenveloping machine, as our average or statistical intelligibility, as the common sense proper to thought (Principle of Sufficient Philosophy) *Identity-of the last-instance* is the cause most inadequate to philosophy, because it is itself the cause of inadequation.

2. *The condition of irregularity or of inadequation itself, the "Other" proper*: to locate and identify the instance of inequality to the continuum (in this case,

the philosophical continuum); this instance no doubt cannot be appropriated by philosophy but, unlike its cause, will necessarily entertain some "relation" to philosophy. It has to obtain the essence or reality of its inadequation from this Identity. This is what we will call *Unilaterality*.

3. These two conditions together eliminate the solution of contemporary philosophies, which consists in employing a *first* Other as an immediate solution. What is at stake is a vicious circle, typical of philosophy: seeking the inequality of two terms, the philosopher confines himself to positing, by simple *petitio principii*, this inequality in itself in the form of an Other, which *already* necessarily has all the traits of philosophical "logic," which is already doublet or fold, mixture of transcendence and the transcendent, "autoposition" in its own way. The "Other" puts the philosophers to work, but it is an argument as lazy as philosophy itself. The sole theoretically rigorous, noncircular solution consists in "locating" or "discovering" an absolutely and not relatively first term. *This term implicates the greatest inequality to philosophy without being itself this inequality. Rather, this inequality will be deduced from it as the strongest relation of inequality to philosophy. Finding its cause or its reality in this term, it no longer forms a circle with philosophy. As a fractality that interrupts continuity without forming a circle with it, without being conditioned by it; and that is generalized in this way.*

4. Geometric fractality is the property of objects that lack uniform, continuous properties, but whose irregularity, however extreme it might be, becomes remarkable, formulable, and quantitatively identifiable. GF is not a property of this type. It is *a property of knowing rather than a property of natural objects*. And this requires it to be purely qualitative and defined vis-à-vis the essentially continuous model that philosophy is. Its concept and its essence have to change. If it is no longer a question of measuring the degree of irregularity of a statement vis-à-vis the average philosophical norm, this is because GF's conditions—its requisites—have an entirely different style and appeal to *sense* and *natural language* rather than to quantification. In particular, we will not say, in the style of philosophies, that the measurable or calculable aspect of irregularity is inessential or belongs to the thing's "phenomenality." Quite simply, it has no meaning, not even a secondary one, in a qualitative and natural-language science ("non-philosophy"). More than qualitative even: because it is thoroughly transcendental by its essence and because, as Identity, it remains entirely outside the quantity-quality couple that is for it one material among others of its representation.

To be sure, knowledge's fractality does not signify that it is *approximate* or that it approximates the real. It is, on the contrary, because knowing is not itself real in the strict sense (Identity) that it is fractal. In general, the idea of approximation must not be conflated with the idea of measuring the degree of an object's fractality. For the approximation may concern the measure of fractality itself. Fractality, in its irreducibility, corresponds instead to the duality Identity-Unilaterality/Material (in this case to philosophy, which is reduced in its claims). The ultraphilosophical generalization of natural fractality rests on all these conditions (Identity and nonresemblance, Unilaterality and not first Other, etc.).

THE CONDITION OF CONSTANCY OR OF SIMILITUDE: IDENTITY-OF-THE-LAST-INSTANCE AS CAUSE OF GF

What is the principle of fractality's generalization, both the cause that produces it and the instance that reproduces it or endows it with its "self-similarity"? A geometric fractal is defined by a certain equality between the condition of irregularity and the condition of constancy or similarity—the equality of two knowledges. A fractal philosophy in the style of Difference is defined by a certain (invertible) hierarchy of irregularity and its constancy or identity. Finally, a generalized, unary, or *real* fractal (neither ontic nor ontological) is defined by the *irreversible priority* (neither simple equality of knowledge nor primacy-hierarchy, but *determination-in-the-last-instance*) of Identity over irregularity. Subject, of course, to the reformulation (*already* carried out) of the experience of Identity and its "relation" (Identity-*of*. . .) to fragmentation. Identity, the condition of constancy or internal similitude, must receive its full sense if it wants to be the cause of fractality and no longer simply one of its given conditions. No longer ontic or ontological Identity (transcendent in both cases), but radically immanent or transcendental. Philosophy can explain on its own a certain internal similitude as well as a certain fractality. Here we are dealing with something else: "internal similitude" as essence or cause (of the absolutely nonmetaphysical type) and no longer as simple property of an imposed inequality or irregularity, which is already given in the World.

The *first* condition of the fractals of knowing is Identity: not any identity whatsoever, not the Identity that would be alienated in fractality itself while effacing it, but the one that remains "in" itself, as it is, at the very heart of its effectivity. Thus: *a cause that cannot be exchanged with its effect*, but a cause that can be located as such *at the very heart of this effect*. Identity is "lived" and conserved as such, unobjectivated, at the heart of a couple of contraries for example, without being itself affected by the play, the exchanges of this couple. A cause is indeed at issue: it produces a single type of effect, its own identity, but an effect that is differentiated according to the material in the midst of which it acts, an effect that constitutes fractality. But it is also a transcendental cause: both by its radical immanence and by the relation of conditioning (a very special relation, since it is "of-the-last-instance") it entertains with what serves it as "experience" or "manifold" (in this case, mainly the philosophical Decision) and in which it continues to be experienced without any change or becoming of its essence.

Only Identity as it is of-the-last-instance is reproduced as rigorously identical *through . . . in* or *directly in . . .* as well as *despite . . .* the permanent variations of content. It is not "reproduced" the Same, but subsists as it is in what is reproduced, without thereby giving place to a simple resemblance or similitude, to an analogy or a univocity, or to a variance of variance— that is, to philosophy's representative generalities. Not only is it the sole Identity that knows itself as this Identity, and is not modified in its essence by its own knowing or involved in a becoming, but even as cause in the midst of its effect of unilaterality or of fractality it is not undermined by this fractality and can be manifested (a second time) as what it is (the first time or in itself).

GF is more than the self-similarity of any system drawn from nature. The system's traits of irregularity and of interruption will not be simply similar or alike, give or take differences in scale or content. It is not simply a matter of a "family resemblance" or the objective appearance of the identity of Difference. There is a genuine *identity*-of-the-last-instance of reproduction, but it does not sink in this reproduction, and a "total" instance, "Being" or "Full Body," does not fall back on irregularity, absorbing or even marginalizing it. The fractal must be opposed to the "different" and to the "molecular"—and not only to them, of course (to the "language game" for example . . .). What occurs as self-similar is not the simple identity, since this identity is the One. It is not the Other (of philosophy), the Irregularity or the Inequality itself.

What occurs "in" Identity is the Inequality-to-philosophy, and it occurs as identical, for Identity is the essence of the Unequal (which cannot be violated by the Unequal). The greatest irregularity, the greatest gap, occurs as identical and no longer as the *Same*. This is a crucial distinction in relation to . . . "Difference" and to other neighboring concepts. The product is not the *Same* in which Identity would be mixed with the Unequal; it is *Identity itself, which occurs—rather than returns—as Identity (of) Inequality*. Quasi reproduction, but without return; noncircular dynamic, in which "linearity" does not give place to any "curve."

This cause of GF, which is unknown to both geometry and philosophy, has a decisive effect on fractality itself. GF is an irregularity, an interruption, vis-à-vis philosophical logic, which is apparently the most extended logic. It is a "relation" of inadequation to philosophical intelligibility and to its procedures of continuity. But this definition is vague. What is it that is irreducible to the philosophical Decision? The Theory of Identity teaches us that Identity—and not some instance of the Other à la contemporary philosophies—is the instance that is most inadequate to philosophy; that it is the authentic "Other," precisely because it is in no way an Other that is autoposited or supposed in the vicinity of philosophy; it is an instance that has already radically suspended and undifferentiated philosophy (the PSP). GF rests on a downgrading of the Other in relation to Identity's anteriority or *precession of-the-last-instance*. Identity is not first *in relation to a second*; it determines every other instance to be strictly second or unilateralized in relation to it.

This point is decisive in the struggle against philosophical illusions. In effect, the absolute Indivision, without blending (but without blending because it knows itself to be undivided and knows this in its own mode of Indivision), is *inadequate in principle* (and even more than inadequate) to every division or decision, synthesis or coupling. Simply because it is immanent (to) itself, it cannot by definition be located in the World or in Being, where it nevertheless acts. This *inadequation-to . . .* is thus no longer an essential property, a predicate of the essence of Identity, one that would retransform it into a species of the Other. It is a simple effect of its essence and does not codetermine it. Identity is not of the nature of the Other, and it is for this reason that it can unilateralize or determine the instance of unilateralization for every other given; it is for this reason that it can give to this instance the figure of the Other. The *dual*'s "non"-relation, the "dual" of Identity and philosophy whose authority is already suspended, does not fall in principle in any philosophical distribution or economy of knowledges (geometric knowledges, for example). More than the Other, which always requires a form of coupling, it is "inadequate" to Inadequation itself. The "last-instance" is more "Other" than every Other; it is only "the principle" or the cause of the most radical fractality. Before describing this fractality in its condition of irregularity, it is important to mark the mutation that this type of Identity introduces into fractality.

There are two different types of conservation of fractality in terms of Identity, which serves as its principle. If Identity is given *with* fractality, convertible or reversible with it (Identity is *itself* fractal or affected by irregularity), fractality will only be conserved as effaced, in the form of a continuous yet superior curve (for example: the philosophical Tradition, Destiny, etc.) that necessarily accompanies it and in which it obtains its sense and value. A conservative or reproductive conservation. If Identity is, on the contrary, de jure inherent (to) self without decision or transcendence, it does not risk falling back on fractality, forming a whole or a mixture with it, a tradition or a continuous curve. Fractality occurs or emerges as new everywhere: not new in a preexisting element, in an indeterminate generality or transcendence that would attenuate it, but new because it occurs or emerges "each time" *for the first and only time*. And it emerges from Identity, with which it is not mixed, to which it does not return, rather than from a background, a tradition, or a reserve that would reappropriate it—rather, also, than as a form against a background. It is intrinsically unique each time as well as solitary—and received or lived as such by Identity itself, or "in" Identity, which is not a *subject* behind its act and its product, which is not alienated in them.

THE CONDITION OF IRREGULARITY OR OF INTERRUPTION: UNILATERALITY

Identity-of-the-last-instance is not only what conserves or "reproduces" fractality as identical (to) itself. It is, in the first place, what "produces" fractality; it is its cause. But Identity, as we said, is not itself fractal. We have to add to it an instance of the Other (fractality proper), which is here secondary and no longer primary and determinant as it is in philosophy. We will thus distinguish, on one hand, the "relation," the nonrelation, what we call the "dual" (or the greatest gap, which is not a first gap) of Identity and the philosophical Decision (or of coupling in general, including the fraction). And, on the other hand, the *solution* to this "inadequation," which is greater than every inadequation, a solution that takes the form of *Unilaterality*. The dual is the structure that is expressed or exists through *Unilaterality*, the Other's true scientific concept and GF's concept. The dual is the cause whose effect-of-unilaterality announces it as the cause that it is.

Qualitatively, what does this structure of *Unilaterality* consist of? With Unilaterality, we pass from vague concepts of irregularity, interruption, and fragmentation—which only become exact when quantified—to a qualitatively precise figure that is, moreover, *the kernel or germ of every fractality*. Unilaterality is a radical and oriented asymmetry, the pure irreversibility or the *Uniface*, the nonsystem of a radically open relation, not teleologically closed by an adverse term, because every supposedly adverse or reciprocal term is in reality absolutely pervaded by contingency. Geometric fractals are characterized by an "irregularity" of form, rhythm, figure, structure, i.e., transcendent and reversible properties, an irregularity that is reproduced and repeated, as if the internal similitude could only appear at the end of extreme variation. On the other hand, this irregularity loses here its transcendent figure. It is in its turn "interiorized" and "autonomized" into an instance or a "fractal order."

The other concepts of fractality are in every respect more complex, more derived and transcendent. They confine themselves to interiorizing common sense's and perception's experience of irregularity outside itself (i.e., in the concept) or to geometrizing it: as a nevertheless symmetric structure in the asymmetry; a bilateral or bifacial, reciprocal, even reversible structure, which presupposes that irregularity is *perceived*, looked at, watched over by the observer, inscribed in the transcendence of the World. In contrast, Identity-of-the-last-instance does not leave itself and institutes with the World an absolutely tapered, stretched, infinitely distant or distancing relation. This relation must be effectively traversed rather than surveyed and thus "philosophized." That is why *we distinguish Unilaterality and the fractionary, always bilateral relation and thus the types of fractality that each of them respectively engender.* Identity is more or much less than a homothety with symmetry and displacement. Likewise, Unilaterality absolutely excludes coincidence, intersection, or *double points*. GF's definition is precisely that it contains no reversibility, but rather a pure distancing or a distance without loop. We will describe these phenomenal traits in detail later on.

What is now this fractality's relation to its cause, the fractality that assumes the mode of Unilaterality? It is not "produced" by its cause in the way in which causes in the World or in Being are alienated so as to continually produce their effect. It is a relatively autonomous structure in its species or its quality, distinct from the immanence of Identity. It does not codetermine its cause in return; it is at last manifested, i.e., produced in a radical phenomenal mode, as it is, through Identity-of-the-last-instance. All these traits are implicated in the "determination-in-the-last-instance," which signifies not only that the cause is not lost in its effect, but that it communicates its autonomy to this effect. Accordingly, Identity as cause makes it possible to dispossess the Other (fractality) not only of its traditional bilaterality, but, what amounts to the same thing, of the false illusions that result from its philosophical or first autoposition, and to constitute it into a *relatively autonomous*, yet secondary order of existence. *Fractality ceases to be a "property" of certain transcendent objects, to be projected metaphysically, in order to become a sphere or an order of reality with relative autonomy, the order of knowledge or theory.*

A CHANGE IN THE THEORETICAL TERRAIN: UNIFRACTALS AND BIFRACTALS

Let's return to the condition of identity and its efficacy on fractality proper. This is the occasion to dissolve a few philosophical appearances that GF can provoke.

Consider the "geometric" definitions of "self-similarity" formulated here and there by Mandelbrot. "It is the property of a geometric form in which each part is a reduced image of the whole." *Scaling* or the property of *internal homothety* "is said of a geometric image or a natural object whose parts have the same form or structure as the Whole, except that they are at a different scale." All the variants of a fractal construction have a "scaling" character: "not requiring a new rule at each stage of construction," but "copying the details of the prior stage, which one will have reduced beforehand to a smaller scale." Every part or fragment has the same form as the whole or is a "reduced image" of the whole. This property is thus self-similarity. And when it is not a question of the Whole, it is the argument of *statistics* that takes its place: "When every piece of the coast is, statistically speaking, homothetic to the whole—save for a few details that we choose not to concern ourselves with—the coast will be said to possess an internal homothety . . . [a notion that] leads to measuring the degree of irregularity of curves that satisfy it, through the relative intensity of large and small details, and ultimately through a dimension of homothety."[2]

It is quite clear that, theoretically, these definitions are not very certain, although their coherence is remarkable and they form a system with the idea that the fractal objects constitute *hierarchized* clusters and overclusters—but only *apparently hierarchized*, specifies Mandelbrot. Here there are perhaps only appearances that have a philosophical origin. And the conceptualization in terms of Whole/part can only feed a hermeneutic interpretation that would efface fractality once again. The idea of scale variation is more rigorous and allows us to abandon the circle Whole/part to the illusions of immediate perception, of common sense and of their philosophical extension. The Whole has no proper reality (like the part as part, the *partial object* for example, the fragment . . . that is opposed to it) except for philosophy. This explains why philosophy's fractality—since such a fractality does exist—is fractionary in a transcendental (transcendental/empirical fraction) and not only empirical or arithmetic way. It is a *fuzzy fraction* because of the rejection of the *Identity* between terms and because of *amphibology* or the "*Same*." The fractality of parts pertains to the superior law of the Whole, even when it is the fractality of the part as such and not the fractality of the relation of (dialectical, hermeneutic, etc.) subordination of the part to the Whole. Philosophy is the operation of drowning fractality in the Same, in Difference, if not in the Whole—in any case, in the unity-of-contraries or the Dyad. We oppose to this *fractality of synthesis* or of totality—obtained *through philosophical synthesis* and presupposing the operations of a third agent, an ex machina philosopher who uses it for his own benefit—*a fractality of identity* that will produce *a philosophy of synthesis, which is called* "*non-philosophy*," *instead of a philosophical synthesis*.

Assuming its geometric form can be effectively conceptualized in terms of Whole/parts, self-similarity has to *change its terrain* in order to become an absolutely universal theory; it has to pass from the philosophical terrain of the Identity-Whole to the properly scientific terrain of Identity-of-the-last-instance; from the possible, where it is dissolved in its own reaffirmations, to the real where it gains life, movement, and being. Fractality can only be generalized if it abandons the Whole/part teleology and even the last closure of the *partial* as such, the *fragmented*, the *disseminated* . . . ; it has to abandon the spirit of hierarchy that its concepts carry, explicitly or as a last vestige from which they cannot be emancipated. It has to stop cofounding itself in the fractalized material and therefore in the *mixture* it forms with the material; it has to discover its cause in Identity-of-the-last-instance, the sole "foundation" that cannot efface it. If no Whole or hierarchy exists any longer, no perceptively detectable *clusters* with identifiable contours (whether curved or fuzzy, continuous or angular), no *parts* will exist; not even *partial objects*, which are always also associated with *flows*—and flows are the last avatars of the Whole. The effect of the suspension of the Whole and of the philosophical forms in general is that there are now only absolutely dispersed *Unilateralities*, whose *chaos* is not limited by philosophical teleologies. The *unifractals* of science or of theory must be distinguished in this way from the bilateral or divided fractals, the philosophical *bifractals* obtained by *first* division or distance and that pertain to the law of the Whole or of the Same (of Difference, etc.).

Unlike geometric fractality, GF—because it is immediately chaos—does not support the continuous interpolation, division, and insertion of new irregularities up to the infinitely small through scale variation. This is also what distinguishes it from philosophical multiplicities. These multiplicities are founded on a division, a generally positive division, of course, on a positive distance instead of the simple division of an identity. But this distance continues to integrate with an identity; in any case, it presupposes an identity in a relation of reciprocity, so much so that this distance and this multiple are primary and, paradoxically, are effaced in their primacy or auto-position. By contrast, GF is "generalized" only because it is no longer primary or autopositing, because it flows from an Identity-of-the-last-instance or comes *after* it. This Inequality "in-Identity" or "in-chaos" excludes every process of division or interpolation. It is a structural, static distance, already given in the wake of the Identity with which it is not contemporary and with which does not form a system. Fractal Inequality is inequality-to-the-World or to-philosophy, but this time it is itself "unequal" (to) Identity as well as (to) its cause.

Scientific unifractals are no doubt poorer or more elementary than the philosophical bi or difractals. But they are not poorer as fractals. Poorer in philosophy, they are more acute in science. They are fractals without curvature, beyond curvature, not susceptible to an infinitesimal analysis. We will oppose them from this standpoint to a geometric or scientific-ontic form *that can always more easily than they can*—although not without resistance—let itself be reinterpreted philosophically as a "miraculous line" grounded in the originary confusion (the similitude of the Whole and parts) and can give place to phantasms of the "living line." Geometric fractality easily drifts toward the "serpentine line" of grace (Ravaisson) or toward the most continuous philosophical curve (Eternal Return of the Same, the Möbius strip), whereas GF and chaos are the most insubordinate critique of continuity.

If one is nevertheless committed to "saving" the Whole and the parts, it is ultimately necessary to admit that the terms *image, similitude,* and *resemblance* take on a *radically objective* sense; that such phenomena exist, but are not grounded-effaced in a relation of continuity, belonging, or reflection of the Whole and parts; that these latter are therefore *flattened* outside *every* relation (in particular any relation of hierarchy, of circle, etc.). To be sure, a Whole would exist, but it would not be first and autopositing. It would be nothing more than parts, which no longer have any relation "between" one another (internal and external relations, little matters now); *each has only a relation with the fractal structure that determines them as generalized-fractal object.* We will see further on that terms isolated in this way—deprived of their reciprocal relation, of their synthesis or of their opposition, and of the philosophical mode of these latter, and now having a relation as individuals or terms only to the fractal structure—form what we called a *chaos* and even a *generalized chaos.* And chaos is the only way to suspend the principle of internal relations as well as the principle of external relations and, furthermore, to suspend their amphibological conflict.

Another philosophical appearance must still be dissolved. GF means, of course, that, regardless of the degree or scale of description of a scientific knowledge and thereby of a philosophy, thought never reaches homogeneous and equal parts, but ceaselessly discovers new inequalities. It is, however, not enough to say that *differentiation*, far from diminishing, on the contrary increases with "magnification" and the emergence of new unsuspected details, that there is a power of differentiation of thought that prevents us from reaching the simple or the homogeneous. Philosophy itself can maintain this (for example, when it combats the nihilist leveling of differences and reaffirms them). We are dealing with something other than a continuous differentiation: the impossibility of drowning the Irregularity "in-Identity" within Difference, the most fractal singularity within the continuity of a process, or again the real within the "impossible" or the "incompossible."

In a general way, in other works already, we have not stopped insisting on a new requirement that is unknown to the contemporary philosophers of "multiplicities" or the "inconsistent multiple": the multiple will have no purpose, will have no *real* critical force if it is not accompanied by a duality, a de jure inequality that ends up breaking a priori not only the "metaphysical" style, but the generally unitary and philosophical style. The multiple through difference (Nietzsche) or through inconsistency (Cantor), but also dissemination through the Other or through Difference (Derrida) represent, from our point of view, an ultimate *unitary* normalization of the greatest fractality, which resides in inequality through unilaterality and which, far from crushing the multiple, emancipates it or gives it the space of its efficacy. The philosophies of multiplicities have, as always, only crossed the easy half of the path, and assumed as a limited target "metaphysics" alone, representation alone, and not philosophy itself.

The science of the One is no longer fractionary—neither in the arithmetic nor the philosophical sense. In general, it no longer inserts any relation of inequality between simultaneous terms (the One, Being, being); it places inequalities between nonsimultaneous terms—we can, in fact, call simultaneity the identity that is given with each of the terms *in the mode of a last instance*. Every fraction is a reciprocal relation. It even becomes reversible, as we see in philosophy, which is reduced to its essence of decision and rid of its "rational" objects. There is nothing of this kind, no *relation*—of fraction or otherwise—between the One, Being, and being as structures of science. What we call Determination-in-the-last-instance excludes all *relations* for the benefit of relatively autonomous *orders*, defined precisely by their identity. The only quasi-fractionary moment would lie between Being and being—again: it implicates no reciprocal relation between them.

The weakening of the fractionary relation is not a weakening of fractality. On the contrary: it is its liberation, its letting-be as autonomous *order*. Its generalization passes through the "destruction" of the fractionary, even the destruction of its philosophical or "superior" form. It is a mutation from the

Fraction's terrain to *Identity*'s. Instead of an interiorization to the concept, fractality changes its basis and its principle.

THE PHENOMENAL TRAITS OF FRACTAL OR NON-PHILOSOPHICAL OBJECTIVITY: THE FRACTAL A PRIORI

We have not yet elucidated the undeniably original relation of the fractal structure to its manifold—in this case: the philosophy it unilateralizes or "fractalizes." Before taking up this task, we should examine the internal nature of the structure of Unilaterality itself, fractality's peculiar constitution as an original "ontological" region or domain, as science's sphere of existence.

Sometimes the geometric fractals can be obtained by recursive construction, by indefinite repetition of the same procedure or the same figure. We reproduce a certain number of these elements in a homothetic way, but at ever smaller scales. This procedure leads to an artificial fractal, ultimately to a fractal representation of a natural object that does not perhaps have the same irregularity. This constructivist and operative aspect (interpolation/extrapolation of the structure) presupposes an *intuition* in which the "concept" of fractality is constructed by an idealizing repetition. But even the "natural" fractals that can be identified "with the naked eye" presuppose such an intuition. They are simple figures, and their regularity or their "concept" is indivisible and can be grasped at once. And we know how much Mandelbrot insists on this aspect, with the complementary rejection of formalism and the axiomatic. So the intuition in all its forms (perceptive or a priori) seems to be a de jure condition of every fractality. What does it become in GF? Is the structure of Unilaterality—the most elementary fractality—intuitive, and what sort of intuition is at issue?

Clearly, we are no longer dealing with a geometric and/or perceptive intuition. The description of GF transferred it from this "ontic" (if not empirical) element onto the terrain of nondecisional-(of-)self Identity-of-the-last-instance, whose radical immanence excludes every transcendence. As a result, if GF is still "intuitive," its cause rules out that it can be a matter of a sensible intuition, but also of an equally transcendent and autopositional *intellectual* intuition. In the structure of Unilaterality that GF conditions, not every transcendence nor even every intuition is excluded (since this structure *is* the very transcendence of scientific representation); what is excluded is the *philosophical form of transcendence, its divided/folded form, its doublet or mixture form*. Scientific representation, rethought in its essence and no longer simply induced [*induite*] from this or that local knowledge, is at any rate the dimension of the objectivity, of the theory and of the objects of knowledge it contains. And this objectivity is at once "sensible" and intellectual. But it is no longer by any means certain—quite the contrary—that we are still dealing with the philosophical form of objectivity, of intuition, of sensibility, and of the concept. The cause (of) science manifests or renders accessible the field of theoretical representation, its phenomenon; but it only manifests it under its own conditions. These conditions exclude the most general philosophical form of theory and suspend the doublet form (the form of autopositional faktum) that philosophy spontaneously gives to scientific knowledge. What is the residue, the remainder of this suspension, a remainder by which the essence of theory can be safely characterized? What is this *real kernel* that Identity-of-the-last-instance extracts in some sense from philosophy—the material of its efficacy—while purifying it of its philosophical form? Here are a few of its phenomenal traits.

• It has to be a question of a transcendence, an opening or an exteriority typical of every representation, whatever it may be.

• Like every representation, it has to be considered objective; it is the very dimension of objectivity, in this case in its form as theory or knowledge. "Scientific" fractality is thus related—we do not yet know how—to a manifold, to a material; and in this case it will only be the very form of (theoretical) objectivity of this material. GF is not only an "objective property" of theory—almost in the sense of the property of a "natural" object. It is also the *property of objectivity*, which is that of science and which, having a fractal nature, is thereby absolutely distinguished from the philosophical objectivity with which epistemologies generally conflate it.

• What does the fractality of objectivity itself signify? Obviously it does not mean that the form of any object, its contour, is a rugged, angular, or irregular curve. We are in the transcendental dimension of the essence and of the essence of theory "in general," not in the dimension of perception and of its geometric givens . . . But it means that knowledges' very objectivity, their status as objects-of-knowledge, is intrinsically fractal or uni-lateral, that this status therefore excludes the *continuity* of the doublet-form or the fold-form. An elaborated and validated knowledge—which does not mean *metaphysically certain* and unrectifiable—is a knowledge only insofar as it is affected by an elementary yet de jure irregularity, by a distancing or a trait of the "faraway," whereby it does not re-turn to the real of which it is the knowledge, a distance at the end of which it *occurs* without ever returning to itself, without looping back to itself and thus giving the illusory impression of the real. This irregularity-to-philosophy, which is typical of scientific knowing, can ultimately be said to exist in relation-to-philosophy, to its continuity (always looped or drowned), although we do not yet know its exact relation to this material. But it cannot be said to exist in relation to the real-of-the-last-instance of which it is the knowledge. It can be this unilaterality of knowledge only because it does not measure itself against it; otherwise, it would conflate the real with a supposedly present object that is given in its turn in transcendence. GF is lived and received by the "man of science" as a state of affairs "in itself": there is no science without this affect of theoretical existence as an existence that is not "in the distance" or "distancing" and "neutralizing," but *in*-the-Faraway. The theoretical view is fractal because it is a view "within" the Distant, rather than an activity of distancing and thus of return, as is the philosophical objectivation. The dimension of the Faraway-without-return is a static property, a structural state of affairs of knowledges and does not result from an operation on the real, on Identity. Such is the phenomenal or real kernel of what we call here and there, without understanding their origin and sense, scientific "neutrality" and "objectivity."

• What is the content in reality of this intimately fractal representation, and what reality is at stake, if it is neither a matter of Identity itself nor of philosophy's mixed reality (real possibility, actual virtuality, etc.)? The anti-epistemological interest of this description is to make clear the existence of an *a priori theoretical intuition* as constitutive of knowledge. The fractal dimension, previously described as immanent view "in"-the-Faraway-without-return, "in" the Distance-without-loop (no doubt the scientific form of infinity, science's infinity rather than one of its objects, a mathematical or cosmological object for example), can no longer be reduced to its philosophical form, i.e., to the doublet, the (variously proportioned) mixture of a representation divided into *intuition* and *concept*. We know that we must think it in terms of its real essence or its cause and that this cause is "in-the-last-instance." Hence two fundamental consequences.

On one hand, scientific representation ("theory" in the broad and full sense, neither empiricist nor idealist, as we understand it) is a priori. It is *universal and necessary* (this is the only necessity internal to knowledges; it is not an external necessity, but the "fractal" system of these knowledges, which we will later on redescribe as *chaos*). More universal and necessary than philosophy itself, since its traits lie in Identity and are no longer philosophically divided (necessity divided as a coherence of rules or formal a prioris and as necessity of a given manifold; universality divided as generality and as totality and thus at once complex or mixed, simultaneously impoverished and reaffirmed . . .).

From where does the *theoretical or fractal a priori* obtain this *absolute and undivided* necessity and universality, which render it simpler and more powerful than the philosophical a prioris? From its cause, no doubt, which absolutely precedes the a priori and no longer forms a circle of reciprocal conditioning with it, as is the case in *every* philosophy. The *last-instance*, having here at the same time a transcendental nature—i.e., it is radically immanent but also conditions the object of knowledge—finally determines the a priori without a vicious circle. The conditions of a rigorous, circleless "transcendental deduction" of theory are finally united: the theoretical a priori is "applied" to the manifold of the philosophical material and produces knowledge under the efficacy of its "transcendental cause" (the "relations" of this "application" will be described later on). Thus are liquidated the aporias—the circles—of relations between the a priori and the transcendental, which Kant and neo-Kantians were unable to resolve because they still thought these relations inside the doublet-or-mixture form, as apriorico-transcendental *fold*. There is here a veritable *transcendental deduction of the fractal a priori (of theory)* whose validity or pertinence for the philosophical given is determined under a double condition: first transcendental, the condition of the real or of Identity; then empirical, the condition of its material, namely philosophy, which is fractalizable to the extent that it is first reduced, suspended in its claims by Identity itself. Instead of a supplementary neo-Kantianism, we have a non-Kantian conception of the fractal at the same time as of "non"-Kantianism as fractalization of Kantianism.

• On the other hand, knowledge's specific reality, the existence of scientific representation, is neither an intuition nor a concept nor their philosophical synthesis or difference; it is their identity—a whole other thing than the old "intellectual intuition," which is a philosophical mixture.

Instead of "concept," we prefer to speak of "theory" in this case. This *theo-retical intuition* prevents theory from becoming an abstract and transcendent construction: it has a properly theoretical content (this fractal objectivation, which is itself complex or specified by several moments described earlier in this book and that constitute the form of objectivity), but it also has an a priori content from the manifold, which is furnished to it at the beginning by its philosophical *data* and that we described as *chôra*. So much so that theory is never a simple indeterminate generality, induced or abstracted from a manifold of given objects, nor a simple empty form of the logico-symbolic type. GF is a *thought* (an objectivity), but it is an *intuitive* thought (it has a priori contents and not only forms). Their Identity-of-the-last-instance renders *absolutely indiscernible*—although without turning them into an amphibology, quite the contrary—the form of objectivity (the "intellectual" side) and the content or the manifold (the "intuitive" side) under the species of which theory is concretely given each time. No line of partition, no philosophical decision, can still pass between those that are thus grasped and received "in-One."

We should thus recognize that there is not only a fractality of sensible intuition, of "perceptive" and natural geometry, but more profoundly a *fractal a priori intuition* that is no longer opposed to the theory or is its simple base, that is the very essence of every theory. There is what we could call a *real fractal a priori* (neither formal nor material), which exerts itself in the form of theory and guides the scientific labor of experimentation. It is a question of theory as we have described it and not as epistemologies isolate, limit, and restrict it to an activity of theorization. For theory is obviously not one *operation* among others, terminal and "superior." It is the very essence of science and what distinguishes science from philosophical "thought." It is thus by means of their scientific treatment that philosophical objects will acquire a fractal nature and a *reality* rather than a *real possibility* and thus become "non-philosophical."

   • Let's come back to the problem of intuition. The extraordinary primacy of seeing, of the eye, both of the geometric and of ordinary perception, in the theory of natural fractals is the *sign of a problem* that GF resolves in its own way. What we called fractal objectivity or objectivity-through-unilaterality is quite unusual: if philosophy distinguishes and unites transcendence and the transcendent, the *distance of objectivation . . .* and the *object*, in a mixture, science identifies them absolutely. On one hand, the object-(of-knowledge) is *directly* the theoretical transcendence; it is identical to this transcendence. The object-(of-knowledge) does not float within theoretical transcendence as in a preexisting element, in an abhorrent universal vacuum. It does not add its own particular transcendence to a milieu of universal transcendence that is destined to receive it. In a sense, there is in theory nothing more than this pure movement of infinite, unlimited transcendence with which the objects of knowledge are conflated. But on the other hand, the distance of objectivation—and it is always somewhat distinct from the object according to philosophy, which assumes that this object is partially hidden, in withdrawal or in reserve within a subject or an agent situated behind it—is in science (where precisely the "last-instance" is not a back-world or a back-act) integrally "objectivated" in its turn, i.e., thoroughly visible like the object itself. Hence the hyperobjectivism that science gives the impression of, but that is in no way the sense philosophers give to it—the philosophical sense of an exacerbated *objectivation*. It is rather the destruction of every objectivation/alienation. The dimension of fractal objectivity is more "objective," if possible, than its philosophical equivalent. It is, as it were, *thoroughly objective*, without an act or agent in withdrawal. There is an *identity* (but of-the-last-instance) of the object and its objectivation. It is a phenomenal state of affairs: that of a *static* or structural *objectivity*, without objective things that add their ontic transcendence to transcendence itself, to ontological objectivity, but also without objectivation to divide and double it, as a verso divides and doubles a recto. The "field" of objectivity, or rather the theoretical opening, is static or unconstitutable by its description and can only be described as it is. It is an entity that can resemble a line (for the line itself), but without being a line; a surface (for the surface itself), but without being a surface; a recto (for the recto itself), but without being a verso; a curve without being a curve, etc. This is precisely the more-than-fractionary sense of theory, its generalized-fractal sense, by which it does not coincide with any dimension of geometric figures, nor with the ontological dimensions of the philosophical decision. We have to meditate on these fractal "images" themselves, to penetrate their unilateral structure, in order to enter into the "spirit" of science and not to conflate it with the spirit of philosophy.

• Lastly, since scientific representation is absolutely stretched out without horizon, an opening without ekstatic-horizontal transcendence, it is rigorously *unreflected* or "opaque." It is indeed a matter of an opening, of a theoretical and intuitive a priori, of a *thought*. But in its intimate nature, it is deprived of every structure of reflection, of fold and refold. It is impossible for scientific representation to be a metatheory (in the philosophical and meta-physical sense of the *meta-*, not in its scientific sense) or an overtheory; it is impossible for it to be separated from itself, while remaining in continuity with itself, in order to survey itself. Science puts an absolute end to the philosopher's position of overhanging, mastery, or coextension, the philosopher who has already surveyed/anticipated *theory* (as the subject that must be able to accompany its representations) because he aims from the outset for a *horizon* of teleological closure of theory and loses the patience of knowledge. The exercise of theoretical knowledge is absolutely immanent, even when it is related to its objects or its material, and this immanence of its theorico-experimental criteria forbids it the hubris of anticipation as well as the nostalgia of retention. It makes the process (not continuous, but *indivisible* or *fractal*) of knowledge a contemplation—rather than an operation or a practice—even within operations and scientific practice. But this contemplation is nocturnal, completely internal, without the light of reason, but not without opening. Precisely because the opening is in-the-last-instance in the mode of intrinsically finite Identity, it is also radically open, without double point or loop, without end or telos; it is even "infinite." This opening is not illuminated in an anticipative way from a horizon that would hang over and close it. It is necessary, not to traverse it (it does not exist prior to thought as an object would, even if it is static and already-open: thought does not exist prior to itself), but to effectuate it *each time* from "objective givens" and "phenomena," provided in this case by philosophy.

Thought's fractality does not thus correspond to any fold or doublet, to any reflection, which are instead processes with continuous curvature. It neither reflects nor posits itself; it remains "suspended" to its cause. *Universality* is a nonangular fractality, without the—always bilateral—angle or fold of reflection. This hyperobjectivity of representation, which follows from Identity-of-the-last-instance, allows us to give a new sense to the affinity that generally exists between the fractal and the artificially produced synthetic images, the computer generated images, or computer graphics, etc. For, more profoundly, if there is an objectivity without objectivation, without object or subject, without poles of identification, without transcendent rules of organization, not only is every theory from the outset a fractal image by nature (although nonperceptive, nonworldly), but its fractality suggests that it is *absolutely* objective (and not semiobjective, semiobjectivating); that it is "flat" or "stretched out," without any thickness or profundity (save for a "surface" one) and can be produced or reproduced at will without requiring reflexive or philosophical conditions, which would separate it from what it can do, which would destroy its Identity-of-the-last-instance and would give it the "depth" of mixture or doublet, the "flesh" of perception. Computer simulation, with the help of mathematical algorithms, of some natural forms that have already been decoded as fractals, cannot be of *direct* and practical interest to us here. But in the application of the science of Identities to the figures of philosophy, we can clearly admit that if nondecisional Identity (of) self, the "last instance," is a sort of *transcendental automaton*, then scientific theory is for its part a sort of *transcendental computer*, which thinks, but in an absolutely unreflected way (and which does not content itself for that reason with calculating); which constantly realizes, through the knowledges it produces, artificial or synthetic images of philosophy and its statements, which are its own "nature," the "natural forms" it can simulate. Here scientific theory contains a priori the conditions of philosophy's fractal decryption and does not need to search for them in an external organon (for example, in a mathematization or a logicization of philosophical statements and operations). GF does not have a mathematical, but rather a transcendental origin, an origin internal to every scientific practice. It applies, as a result, to natural and philosophical language. This problem will be taken up again in connection with "Artificial Philosophy."

THE CONDITION OF CONTINUITY: PHILOSOPHY

Before describing the fractal structure more precisely, we have to reintroduce the fractalized object itself: philosophy.

In effect, what is traditionally called "thinking"?

What is called "thinking a phenomenon" can be reduced to a few invariant operations that define what is called "philosophy" or the "philosophical style." Whether it is a matter of a "regional" (economic, aesthetic, sociological, semiological, ethical, etc.) or "fundamental" and expressly philosophical interpretation, the minimum gesture could be the following: *to any phenomenon = X that we desire to interpret, we associate a continuous plane or space—a universal—of diverse operations, rules, and objects.* These latter can be specified as economic, aesthetic, etc.—little matters. What counts is the *association* of a universal space to the singular phenomenon, a space that can be partially controlled and that serves to interpret and control the singularity under conditions of continuity.

What does "to associate" mean here? In what way is it a matter of a superior "associationism" proper to the philosophical style? We suspect that it cannot be a matter of an indifferent juxtaposition and that, moreover, a dialectic would be only one possible version of the matter. Association means that the singularity and the universal space of control intimately co-belong and cannot be defined outside each other, that there is between them no neutral zone or indifferent field of reality. Philosophy is the theory and practice of singularity as mixture or blending of contraries, nothing else. This is an ultimate invariant to which *every* philosophy can be brought through a supplementary interpretation.

To interpret, to comment, to dialectize, but also to deconstruct, all these operations—their reciprocal difference matters little, it is now irrelevant—have something in common: they respond to the invariants of the philosophical operation, of which they are modes. And this operation is always an idealized and transcendental, nonscientific geometry. More exactly: *at once*—since it is a double or divided operation—*a transcendental decision and a transcendental topology.* There is no event = X to which a universal space of presentation, of resonance, or of interpretation is not *connected*; no individual without an "associated milieu" that outlines series of possible interpretations. Thought rigorously, the relation would thus be that of a "difference," whatever the way in which this difference is then varied. "Association" thus implies several invariant phenomena:

• The singularity belongs to the plane; it is a point of universal space, which can be defined by its coordinates or its properties. At the same time, however, it does not belong to it or not entirely; by a more empirical side, it is given as falling outside the plane. Singularity is, by definition, *divided* by the plane itself into an empirico-multiple, empirico-singular side and an ideal-universal side. This division of the phenomenon, its splitting, is not an inevitable accident or evil; it is the first operation of every thought that obeys the philosophical paradigm, a practical operation of dyadic scission or decision. Singularity is then distributed on two sides, empirical and ideal.

Since its unitary concept is divided in this way, it involves it in a becoming of synthesis, of reconciliation, of production, of conciliation, etc. Singularity is never given as such, but, on one hand, it is *supposed to be* given, affected by transcendence or divided and, on the other, it is the end of a process, the outcome of a "concrete becoming." These are the real's philosophical avatars . . .

• The second trait, after division, is precisely the trait of identity or identification. By its ideal side, the one inscribed in the space of control, the singularity is affected by an already mapped-out future: to finish its identification with the totality of the plane; to become adequate to its expanse; to be completely interpreted and assimilated by the system of parameters (economic, aesthetic, semiological, etc.). Thought rigorously, this operation of interpretation or of reading of the phenomenon amounts to refolding the space onto the singularity and to *stretching out* the singularity to unlimited, though always equally finite dimensions of the plane. After the singularity's division (a nonempirical division but in the dimension of the transcendental), there comes its *doublage*, doubling, and redoubling in the form of a simultaneous becoming-universal-and-concrete. The duality of singularity was "in itself"; now it is "for itself" or manifests itself as such.

The singularity is thus given twice—this is its philosophical regime. Once as divided from itself or underdetermined and a second time, but it is the "same," as doubled on itself or overdetermined. Determination is knowingly weakened so that it can be reinforced. To philosophize is to multiply the doublings and the envelopes, to accumulate the representations and the control over singularity. All these interpretations, which "have to be able to accompany" the singularity, begin by dissolving its reality in order to present themselves in their turn not only as an overdetermination, but as a constitutive codetermination of its new reality. Semiological, sociological, psychoanalytic . . . spaces—and this doubling of all doublings, the ontological space or philosophy—conceal singularity and reality by claiming to manifest and constitute them.

Reduced to its minimal "eidetic content," this is what here and there we unwittingly call thinking. This "continuous" paradigm belongs to philosophy. Here nothing apparently allows us to speak of a philosophical massacre of singularity or of determination in the name of mixtures. And yet it is in this sense or this sentiment that we are engaged. The complexity of philosophical operations and the very great simplicity of this schema cannot be raised as objections in this case. The complexity is simply the development of the simplicity and thus proves nothing. On the other hand, in the apprehension of singularity as mixture by the philosophical paradigm, this forced labor, this interior and exterior decision, this operative (even if transcendental) side of philosophy—its "technology" side—leaves us scientifically unsatisfied. Transcendental technology to be sure, in that it bears on the whole or the essence of the real and not only on a part of the object. But if philosophy is a transcendental technology of continuity, in what sense does still respect singularities?

The problem, as we have said, is to know whether the real, in order to be what it is, needs to be associated with a possible, to be divided in itself and stretched in the form of a universal space or a continuous curve; or else, whether it not diminished by this amphibology with the possible. Philosophers, who are often all too hasty (some even integrated the haste into the essence of being), will admit without much hesitancy that the multiplication of fundamental principles (the One, Being *and* the Other; the Same *and* the Other; Identity *and* Difference), that the dyad of principles is a source of enrichment, of complexity and of rigor. This is a tragic illusion: the unitary multiplication of the principles of Being engenders their inter-inhibition, as if the possibilization of the real and the realization of the possible were equivalent, through their chiasmus and their accumulation, to a real that would have already been determined *in* itself. This philosophical procedure of division and of doubling is, in the same measure, the scarcity of interimpediments between principles, the war at the heart of the real. Within philosophy's framework, sufficient determination never meant anything other than scarcity and war. The sufficiency of reason is another name for the blending of principles. And the underside of this blending, the ground on which complexity rests, can only be the impossible partition, the transcendental rarity that sows violence throughout the entire real. By definition, the real is enough; it is not rare, this problem does not arise for it. On the condition, however, that we first dissolve its amphibology with the false "sufficiency" of reason. If philosophy is a practice of weakening the real (it divides the real, refolds it on itself; it stretches, doubles, and envelops the singularity, represses or resists it), the extreme tip of this resistance to the Determined is precisely what is presented in the history of thought as the *Principle of Sufficient Determination*. A reevaluation of the reality and materiality of singularities should pass through the unconditional suspension of this principle, which condenses all the philosophical equivocations.

GF'S AUTONOMY: ITS CONSTITUTIVE ORDERS

We can now begin to describe the relations between the fractal structure and the fractalized object. We progressively make our way toward the scientific concept of "generalized chaos."

We have defined GF by internal and specific criteria: by its essence (Identity) and by its specific structure (Unilaterality). It is remarkable that we do not need to define it as "relative-to . . . ," to the material it fractalizes. Here the Other, fractality, unlike what happens in philosophy, is relatively autonomous with regard to Identity, but it is *sui generis*, independent in its essence from the continuous curve it breaks. This is the essential step in the conquest of a GF. And yet this Other is clearly not without relation-to philosophy, even if it is without relation-to . . . Identity. This is the problem of the condition of continuity, which has become secondary here and which we have had the occasion to examine vis-à-vis philosophy.

In a certain way, GF is situated "between" Identity (of) self and the given of the philosophical Decision. It is thus true that in a first, very external, approximation it also has the nature of an intermediate being: between the continuous, more or less hesitant curve of philosophy and the Identity that is so "Other" it precedes philosophy absolutely and without alterity. If fractality exists, it is situated "between" nondecisional Identity (of) self and the philosophical Decision. But if the exact meaning of this "in-between" can no longer, by definition, be philosophical or associated with a philosophy, if it depends for its reality on Identity alone, then it must be particularly elucidated.

If philosophy's fundamental space has a fractionary dimension 2/1 (subject to the aforementioned reservations), the fractal at issue will not confine itself to filling this space once again, nor even to overflowing it. For philosophy is already this line of thought, which, by dint of folds, points of catastrophe or transmutation, retrogressions and turns, can exceed this space and approach the surface. GF, moreover, cannot be posited in relation to the philosophical space taken as unit of measure. It does not even extend beyond this space in the manner of an Other, deconstructions' Other. It comes from elsewhere—from the Identity that is not an "elsewhere"—and it comes "in front" of philosophy, interrupting it radically and without negotiation. It is GF that defines a new space of thought, a quasi-space, that includes in an unequal mode the philosophical 2/1 space and its restrained fractality. The meaning of fractal inequality changes. It is not the fractal that is unequal and relative to the norm. On one hand, it is real or "identical" as inequality. It becomes a fractality de jure—more than de jure and other than a "property." It is "in self" [*en soi*] insofar as it is identical-in-the-last-instance, i.e., in itself [*en elle-même*] (where Identity acts) rather than far from itself (the "last-instance" is not the transcendence of a distant cause, quite the contrary). On the other hand—here is the consequence—it is instead the norm of philosophical continuity that becomes unequal to it, absolutely and irrevocably unequal because it is unilateralized by GF. Fractality is not itself fractal (not doublet or fold); rather, it fractalizes philosophy. Thus this inequality in philosophy is not a property of philosophy itself. It is the Inequality/Unilaterality born from the Identity-of-the-last-instance that is imposed on philosophy and "renders" or "makes" it unequal, radically fragments or interrupts it. It is the norm that is pervaded by fractality, rather than fractality that is decided *in-relation-to* the norm—but this is because fractality exists in some sense "in itself" and forms an autonomous sphere of existence. It is clearly the determinant-of-the-last-instance cause that requires this new distribution of unilaterality. The term *fractality* should be maintained because this apparently intermediate sphere is the Other. GF is no longer the Other-of . . . Identity, but the Other-of . . . philosophical continuity; it affects this continuity without becoming relative to it in its turn.

In relation to GF, philosophy's fractionary space in a transcendental mode appears as a "whole," for example, as a circle or a continuous curve. Even if philosophy's *line* is not simply or only continuous, if it is semicontinuous, semidiscrete, it remains necessarily a curve, whether it is topologically drowned or not, and it is "in relation" to this state of thought that GF appears as something other than a curve or a fraction (or a doublet or a mixture). If we are committed to keeping the term *fraction,* not only should we call it transcendental, as we do for philosophy, but we should modify the sense and scope of the fractionary *bar* and stop positing it as primary. Instead, we should posit as primary the *Identity (of) Inequality* and assume that there is really no continuous bar that unifies the present parts, that Identity-of-the-last-instance is not a "part(y)" in a game or a division.

The "fractal dimension" thus continues to exist, even as generalized. It defines a degree or, more exactly, a type of irregularity:

• unquantifiable . . . In ontic sciences, the degree of irregularity or fragmentation of a curve can be quantified. Already in philosophy this is no longer possible or it is a matter of an ideal quantification. In non-philosophy there is no longer any possible quantification. The greatest fractality is qualitative and lived immanentally. And yet, as the science of the One that is a science of philosophy, of sense and of language-in-philosophy, it is indeed scientific. As we understand it, *first* science, which is homogeneous to other sciences, is a *transcendental mathesis*;

• without autonomy vis-à-vis Identity itself (not "first" like geometric or geometrico-philosophical fractality). Identities in themselves are not fractal and will never *become* so on pain of falling back in their turn (as a *continuous* Whole or a *continuous* Tradition) on fractality. But they are the condition of reality, the cause of fractal structures that form the tissue of the theoretical representation (of) these Identities or (of) the real. But even if fractality needs Identity, it is a richer concept than this Identity; it contains a supplementary "dimension," the fractal instance proper that enjoys a *relative autonomy*. If philosophy makes no irreversible and stable distinction between Identities and fractality, which affects them in return, science is founded on this distinction and, furthermore, on the cause that renders it necessary and grants fractality a relative autonomy;

• autonomous on the other hand, in its essence at least, vis-à-vis the philosophical manifold it fractalizes;

• partially dependent, from the standpoint of its existence, of its object or of its material; it represents in itself a strict and infinite—"chaotic"— fractality (albeit without a procedure of interpolation or division), an essentially perfect fractality. It obtains this relative autonomy from its cause and it is with this autonomy that it occurs as radically self-similar.

GF is not an operation or the outcome of an operation on a prior or related identity: division, distanciation or differentiation, recursive interpolation, etc. Identity-of-the-last-instance is not prior—given in a presence— or associated with such an operation. It is given "in" itself outside every presence, in such a way that if it is followed by an effect, this will not be an operation or an act, but a phenomenal and static state of affairs to be described rather than produced. This state of affairs is fractality as unilaterality. And fractality, unable to form a system *with* Identity, which lets it be, is abandoned to itself, to the distance without closure that it is; opening without horizon or project, without loop or node. It is impossible to inter-polate or extrapolate voids and additions, lacks and supplements, from an Identity that is not inscribed in Being or Presence, in Transcendence, an Identity that remains only "in" itself. Unlike geometric fractality, Identity is no longer the fractalized natural support, it is the *cause* of fractality; it frees it to itself, to its more pure state without limiting or inhibiting it: fractal-ity *in itself* or as such. Conversely, the fractalized support or material (phi-losophy) is less an identity than the mixture of identity and of difference (infrafractal or semifractal mixture) that constitutes philosophy's essence. Restrained fractality belongs to the great sphere of technological objects, of objects obtained by division. Generalized fractality belongs to the imma-nent or indivisible phenomenal sphere and can at most be described with-out being realized. The first has to do with *properties* (of objects givens in transcendence), the second with *essences* that, by definition, can never be detected or identified in the World, but belong to the most immanent expe-rience of thought.

THE FRACTALIZED MATERIAL: OCCASIONALIST CONCEPTION OF THE FRACTAL OBJECT

What occurs as irregularity identical (to) self in-the-last-instance is Unilat-erality's qualitative or specific order, its structure insofar as it is manifested in the mode of Identity. It occurs, as radically identical, let's say *directly and immediately* in the philosophical manifold that accompanies it as well as *against* it. As fractality of-the-last-instance, its relation to the manifold is complex. It cannot be identified as a "natural" property of this manifold (philosophy), it occurs elsewhere, and yet it forms a radical irregularity that does not work *against* its material *except directly* on this material, if we may phrase it this way. We will not therefore look for a "contradiction" to media-tize or resolve between the manifold of philosophy and the fractality that structures it according to this strange relation of the-last-instance. There is nothing to mediatize; there are no longer any "contraries." Fractality is indeed the fractality *of* philosophy, but insofar—at any rate—as the *of* no longer signifies a belonging to philosophy, a "natural" and "originary" prop-erty of philosophy, but a coming (from) of fractality itself as an autonomous order *to* and *toward* philosophy, the advent of fractality "in front" of philosophy, which is really affected by it, but only from the perspective of this fractal order itself or of the Identity that subtends it.

GF is a causality that, like Identity's causality but perhaps less radically, affects or transforms an object without this object conditioning the action in its essence, without its own action surpassing the action of an occasion. We have to be more precise. We have not set out in detail—this was done in the previous chapters—the exact role of this occasion. But GF, the GF of science over philosophy or of philosophy when it is "seen-in-science," presupposes an "occasionalist" conception of philosophy's role and thus of the fractal "object." Philosophy, its manifold at least (PSP is suspended)—not only its invariant representational content (its statements), but its structure-of-decision—is what covers or recovers the fractal structure, what incarnates it: its role is limited to that. GF is no longer the reciprocal system of an external operation or figure and of a transcendent manifold that codetermines it in an equal way (the fractality *of* a figure, of a curvature *of* object, a fractality specific to them). It is a relatively autonomous order, *specified* by this manifold, but not *specific* to it; it no longer presupposes the action of this manifold for the essence of fractality itself, but only for its *material*.

If the "fractal" object—qua fractalized material or support—does not act on the essence of the fractal structure, if it is only a cause in the mode of an occasion, the fractal structure on the other hand acts on it and determines through it a fractal manifold whose concept will be crucial for the definition of chaos.

If Identity is not alienated in fractality, fractality will not be alienated in its material. It remains universal each time it occurs as unique—thus "always." But, or because of this, it affects its materials in each of their points or sites, since it *determines* them as *materials*. This is not the case of geometric fractality, sandwiched between two generalities: its homothetic identity, which is not absolute or absolutely determined on the plane of the real, because it is merely one property or one knowledge among others; and its geometric or physical materials, which it is incapable of cutting through from end to end and remain partially outside fractality. In such a way that some play or some indifference, some nonfractality, envelops fractality and attenuates its vigor. Identity, whatever its place or the object at issue, must be seen "in" itself (what we elsewhere call the vision-in-One or the One-in-One) and must thus serve as an unbreakable guiding thread, so that fractality can—without deteriorating—penetrate its material and determine it precisely as *fractal manifold*. Determination-in-the-last-instance is neither a totalization nor the smoothing of a curve. It safeguards the fractal manifold from every reappropriation by philosophy's continuous procedures. Only the most solitary, the least alienated Identity can give the *fractal order* its identity-of-instance and assert it with respect to philosophy and against it, against its autointerpretation.

THE CONDITION OF PHILOSOPHICAL SCALE-AND-DECISION VARIATION: THE CONSERVATION OF FRACTALITY

Complete fractality has a double aspect. It is not reduced to Identity even if it manifests Identity as it is; but it contains the side of Unilaterality as a radical irregularity. Unilaterality in its turn has two aspects: by one of its sides, it is this Inequality in principle, which does not cease to reproduce itself as identical-in-the-last-instance without erasing itself as a result; by its other side, it does not stop varying its philosophical content. This irregularity-to-philosophy, this distance-from-the-World, reappears identically through variations, those of the types of the philosophical decision, those of the scales at which the content or the details of this or that decision is apprehended. The *condition of variation* will thus change in relation to geometry.

On one hand, it is complexified in decision-and-scale variations proper. The manifold of natural language, which incarnates every philosophical decision, corresponds to the details of figures that can be accessed through rescaling or variations in magnitude. But the change in the philosophical decision corresponds to a variation in the region or domain of "natural" fractality. Consider a vague definition of the fractal as formed of similar structures whose elements are ever smaller. Just as the internal similitude is insufficient and must go up to *identity*, so on the side of dimension the scale-and-magnitude variations must go, in the objects of knowing, up to qualitative variations of the philosophical decision and not only of the order of magnitude in the "detailed" analysis of a philosophy. More precisely, the manifold to be fractalized is not so much objects or categories as three qualitative kinds of manifolds: 1. natural language or the invariant representational content of a particular philosophy, 2. the historico-systematic manifold of decisions, 3. finally, the manifold of structural moments of a philosophical decision as such. It is first and foremost in relation to these constitutive manifolds of the philosophical, and above all in relation to the last one, that fractality asserts at once its irregularity and this irregularity's order or constancy.

On the other hand, the condition of variation no longer plays the same role as in fractality's geometric or philosophical forms in which it coconstitutes and codetermines this fractality. It becomes contingent. This does not mean that it is not necessary, but only that it is not necessary to the structure and essence of fractality. The structure of Identity-Unilaterality is a sort of *homothety*, but it is absolutely internal. It is no longer legible in the World; it can only be read in thought insofar as its cause is absolutely immanent: a transcendental homothety. It is indeed an *invariant*, but, unlike philosophy's semifractal invariants, it is entirely immanent by its cause and not given *at once* in the two supposedly equal modes of Immanence and of Transcendence. Far from having a statistical side like the philosophical, or at least an objective appearance of identity or "sameness," far from being codetermined in its invariant nature by the manifold, it is a strict self-similarity that is not itself modified by the emergence of a new manifold, that does not give place to the appearance of a Same, but that on the other hand ceaselessly gives a new figure to this manifold. The change in philosophical decisions or in scales within the description of a decision does not modify fractality itself. It does not produce a simple "self-similarity," a "family resemblance" or a continuous "curve," an "allure" or a "style," in which case the manifold would "fuse" with the fractal structure, as in philosophy where the very essence of fractality is continually modified and thus effaced or limited by its companionship with Identity.

GF is a constant and a specific constant *for* each text or statement produced as "fractal." One constant is distinct from another inasmuch as it is specified by the manifold of text-materials. This manifold is in fact included in the fractal structure as what is fractalized: it does not determine the essence of fractality. On the other hand, it intervenes in fractality for some precise functions that are variation factors of fractality itself. It is not a matter of invariants or of variance-of-variance, where *identity* would be certainly invariant, yet mixed or combined with the philosophical forms or contents, and would be deformed with them. There is fractality because Identity remains what it is and is not transformed with its contents.

This nonconservative "conservation" of radical Identity, which is everywhere the same in different philosophical milieus, produces fractal structures of a new type. *The variations in philosophical decisions or contents, increasingly analyzed and diversified contents*, reveal, or allow to be manifested, not the "same" structure, but rather the *identity* (of) a structure. If singularities and differences are founded on the *Same*, on the conservation of variations, gaps, or differences, generalized fractality is founded on the conservation of strict Identity. The details vary according to the chosen philosophical scale or decision (a given philosophy does not reveal the same content from the perspective of this or that other philosophy), but each has an identical structure or internal form. The main property—in sciences or "natural" fractal objects—is internal "resemblance" or "similitude," *but in the domain of the objects of science, then of philosophy, the property must go up to Identity*. For if it were a question of a simple resemblance or similitude, philosophy would be powerful enough to explain it on its own. But we are searching for a fractality that cannot be reduced to philosophy's laws of continuity.

Furthermore, this contingency of decisions is necessary for the *exercise* of fractality. We will show that it has to aim—but precisely *without objectivating it, thus without letting itself be codetermined by it*—for a certain manifold, supplied in this case by philosophy itself. The variations in magnitude or in position-and-decision are no longer necessary for fractality itself. They are relevant only for the construction or manifestation of a concrete fractal object in which they intervene as factors of transcendent specification and incarnation, as quasi-philosophical modes of existence of irregularity's a priori structure. What is modified is the manifold itself, not only through its external decision-and-scale variations, but more profoundly through the fractal structure that is imposed on it and that is *identical* and not only *similar, analogous*, etc. As to this structure it does not vary insofar as it is thought and received from its cause, from its identity-of-the-last-instance, *directly* from the manifold, but only inasmuch as this manifold specifies and over-determines it.

The introduction of radical contingency—more than the "aleatory"—signifies philosophies' indifference. What we have elsewhere called *the principle of equivalence of philosophical decisions* receives its full sense here. Its very function is to introduce an element of chance, even of "chaos," into the collection of objects or to *randomize* it. The material is *randomized* in the form of an equivalence—which is required by fractality itself and is not arbitrary or imposed from the outside—of decisions that organize the language that incarnates them. In this way the initial idea of fractals is radicalized, according to which there is no difference of nature between extreme variations in time and space, but rather the continuity of an irregularity. It is now the variations themselves, and not their difference in time and in region, that are contingent in relation to the fractal structure. We said that fractal Inequality emerges as Identity without returning as Same; that is to say: this type of invariance finds its reason in Identity as cause rather than in variations in the degree of resolution or in scale expansions.

This change in fractality's terrain forces us now to distinguish the rigorously "self-similar" fractals, those of science itself, from philosophy's, whose self-similarity is merely "statistical" or "average." For philosophy itself, its own fractality is unstable. It become constant and forms the specific order of philosophical disorder only for a science of philosophy. However, the constancy of the fractal mechanism no longer means, as we saw, an analogy, a similitude, a resemblance, or a univocity— philosophy's nihilist, be it "superior" and counternihilist, boredom. From one decision to the other and from one region to the other of the same decision, fractality is *identical* only *in-the-last-instance.*

Thus fractality can be truly generalized only if its cause, the Identity that remains in itself, is not alienated in it; and if, correlatively, the condition-of-variation becomes more contingent than it is in philosophy, but also in geometry where "chance" and the "aleatory" fulfill this function. Identity remains what it is without moving to the Same, without drowning in knowing or in nature. And the manifold of variations becomes for its part absolutely contingent and plays its role in the form of this very contingency. The two sides of fractality are emancipated from their ongoing identification, from their philosophical becoming.

FRACTAL INTENTIONALITY: INTERFACE AND UNIFACE

A last element—not the mixed form itself as a philosophical relationship or relation between two given terms, but the relation of this mixed form to its manifold—also undergoes the effect of Identity and enters into fractality. What does it become in order to be the new relation "of" these new terms?

On one hand, the mixed form (we analyzed it in previous chapters) lets itself be decomposed into a certain number of a priori functions or structures, those of the philosophical Decision or the procedures of its continuity. These structures lose, under the effect of Identity, the mixed form itself, their autoreflection (in this way reflection can no longer be said of itself). And they become *non-philosophical a prioris*, nondecisional and nonpositional (of) self.

On the other hand, these a prioris, which are now proper to the fractal structure, are, as such, endowed with an intentionality (a priori *of . . . for . . .*), an aiming for experience or for the fractalized manifold. But this intentionality can, in fact, no longer be that of consciousness or Being, which has the mixture/doublet form and objectivates its "object." This is the solution to the previous problem: *the relation of GF—of Unilaterality—to the object "philosophy," a relation of which we said that it was an "immediate presence to . . . but against . . ." this object, is nothing other than this nonmixed, nonfolded intentionality, this simple "aim" and this simple "of " that no longer objectivate that to which they are related.*

It was a question of resolving the problem of geometry's application to physics or of the validity of the former for the latter. This problem is generalized here in the following form: *how can GF be the fractality "of " the object "philosophy"?* How can first science become the science "of" philosophy? The solution has been established. Unilaterality "in-Identity" is indeed lived as close as possible to this manifold; it is oriented on it without being essentially conditioned by it. It has a transcendental function. It is not a transcendent and abstract form imposed on the manifold; it is not the "autoposited" or "supposed" Other, which alters this manifold of philosophy from the outside and arbitrarily. It is the One-Other, the Other "in-One," and this explains the otherwise inexplicable precession of the Other in relation to this manifold.

At bottom, what we have elaborated, at least in principle, is a fractal a priori—an *a priori fractality*. It is thus also the existence of a genuine *fractal intentionality* in the following double sense:

1/ GF is not a simple transcendent property, inert and given with the World, whether this property is that of a "natural" object or a prior knowledge that is simply reified in "nature." It is endowed with a genuine *immanent intentionality* of a special type, which does not objectivate its manifold.

Having a radical cause that determines its nature, it is destined by this cause to fractalize its object. We must thus distinguish not only the *cause* of fractality, but fractality itself and the object it fractalizes.

2/ Conversely, the intentionality of fractality is itself fractal and its concept is obtained in this case, on the theoretical plane, through the fractalization of intentionality's traditional phenomenological form.

What can this fractal intentionality mean for the neighboring concept of "interface"?

Science is a fractal thought whose essence of radical immanence nevertheless rules out that it take the form—even the metaphorical form—of a *figure* (line, surface, volume) in general. In contrast, philosophy is a *curve* or a line that tends toward a surface, as we saw, and that touches in each of its points on an extraphilosophical manifold. Science is so fractal that it does not take the form even of a semicontinuous curve and does not touch in each of its points on the philosophical as on a supposedly "nonscientific" manifold. In its essence, it is not an interface—at least in the traditional epistemological sense of the term—and does not have an interface with philosophy. But, as science of Identities, it nevertheless uses philosophy primordially. In thought's domain only philosophy and technology are interfaces and semifractalities. And it is science that introduces the most radical fractality at the same time that fractality affects the philosophical curve, limits its power of interface as such, an action for which fractality uses no supplementary interface. In the particular relation it entertains with its "exterior," with the philosophical, it suspends straightaway the philosophical's claims over the real, the functions that philosophy exercises through its structures of interfacing; it reduces the philosophical to the state of materials stripped of every claim over itself and over the real. There are no lines and no surfaces of separation common to science and philosophy, which have essentially different functions. Scientific fractality is not codetermined in its essence by philosophy.

The description could, nevertheless, be more precise. Science is not an interface in the techno-philosophical sense of the term, but we can assume that it represents the real, radical kernel of every possible interface. This is what we call the scientific a priori representation (GF's order or instance) as *nonpositional reflection (of) the real*. The ensemble of a priori structures of scientific representation, and thus of Unilaterality, functions as an absolute interface, a "surface" of monstration or manifestation of scientific knowing, which is infinitely open and is thus not destined to be received, captured by a third person, by a supplementary observer like the philosopher. Knowledge objectively "shows," without any external *subject* to receive and redistribute it. Unilateral interface, if you want, which is in reality a *Uniface*, a structure without bilaterality, not a common frontier or limit of more or less interrupted or inhibited exchanges, but a "surface" *without verso and without loop, without borders or folds*. In this sense science is the experience of an a priori interface. But this interface does not form a system with the philosophical materials it treats.

Conceived in this sense, the interface is entirely fractal. For the fractal-generalized object par excellence is scientific representation. It is not obtained through division—interpolation and extrapolation—of a transcendent irregularity composed of folds or of angles, but through the Identity-of-the-last-instance (of) Unilaterality. So much so that the interface and the most irreducible fractality are identified without remainder in Unilaterality—we can see why fractality no longer takes in this case the ready-made form of the peaks and angles, the triangles, the promontories and bays that made it "natural."

FRACTALITY OF THEORETICAL SPACE-TIME

Identity-of-the-last-instance is indeed an identity "of" contraries, but there is here no aiming for, no intentionality of these contraries by Identity, which does not emerge from itself and is not alienated in what is seen "in" it. If theory itself, the fractal or unilateral objectivity, as we have examined it, *aims* through an intentionality of a particular type for the materials it fractalizes, this structure is ruled out for the cause (of) theory, which does not entertain for its part any *relation* (of transcendence) with this structure. It is the manifestation, the phenomenalization of contraries, but in its own condition of nondecisional immanence (of) self. So much so that a new "economy" of these contraries flows from it, a distribution that is not mixed or unitary-hierarchized. It immanently individuates or "identifies" each term outside its relation to the other (hence the quasi-phenomenological layers, strata, spheres, instances, etc.)—for example, the theoretical and the experimental.

But it does so each time by radically identifying the old "contraries" outside every simple relation of co-belonging or synthesis. Identity of each term liberated from the other? More exactly: liberated from its *philosophical relation* (internal, substantial, dialectical, external, differential, etc.) to this other term, but not from this term. Identity is certainly "present" in each point of the materials or of other instances. But it cannot be reduced to these instances. It is seen "in" itself only, "identified" in an absolute way and not as "X or Y," as identity *of* X or Y. It implies that the philosophical couples of X and of Y cease to be couples, that their terms are adequately recovered, but once as X and another time as Y.

The dual, the greatest inadequation, proceeds as a strict or immediate *identification* of contraries, *but under the sole law of Identity* . . . This identification is immediate, nondialectical for example. It does not take place in the One itself—which remains "in" itself without being affected by this identification—but in an element that is produced as secondary and that renders this immediate identification possible. Yet even if it takes place outside the One, it nonetheless takes place *through* the One and *with* the contraries of the philosophy-or-mixture form. Even if it is not a matter of a dialectical identification, there is a production of a middle term, of a *relatively* autonomous instance: the *a priori theoretical representation* or science. *It is in Transcendence, with it, in its element, not in the One itself, that this identification of contraries takes place and can be called absolute, antiamphibological, the identification that is given at the origin as a material in philosophical transcendence.*

What form does this process of determination-in-the-last-instance take? Just as Difference *is* not 1 = 2 but the becoming 1 ↔ 2, and just as becoming is mediation become universal, so Identity—which *is* immediately the contraries (certainly in-the-last-instance)—nevertheless needs . . . or produces a third instance (also a pure *becoming*): theoretical representation. With this difference—since there has to be one: *this becoming-(of)-knowledge is not the becoming of the real*, of Identity itself, and does not involve it in any way. It is thus an absolutely straight or unlimited Becoming, without loop or topology; without returning to itself—without the Eternal Return of the same. *Nothing returns, not even difference, and there is even no "difference" to return.* Becoming through and through, *already* the Open or the Faraway, the always-already-Faraway as such; Unilateralization and never the *bilaterality* of Difference. Every survey-return, every reflexivity or semireflexivity of the type "Difference" is excluded.

The two contraries must "be" 1, must be-identical strictly, 1–1 and not 1–2. They must undergo the 1 that is nothing but 1, which is not at the same time 2, and which alone will determine them as 1. Here it is not the 2 of Difference that is autoposited and reflected as 1 (thus always still as 2) in itself. No: the 2 must be immediately 1 that is nothing but 1; it must undergo the form of Identity. This is possible only if it engaged in a *becoming* (every becoming is always a "synthesis" of contraries), but absolutely stretched out = 1, with dimension-1, deprived of every autopositional, carnal, reflexive thickness. A few nuances have to be introduced into the preceding description.

On one hand, it is tricky to speak of *identification of contraries* without adding anything more. This process would be that of Difference, in which A is identified not with B, but also with the becoming-of-B. It is thus a matter of partial identification, of becomings rather than of objects. Here, in contrast, Identity remains in itself and excludes every identification (even partial identifications). The contraries are not grasped through their reciprocal or reversible *identification*, but through their *identity* (of-the-last-instance), which is not engaged in their affairs. It cannot thus be a question of a simple identification of contraries among themselves, but of each of the terms as noncontrary, without contrariety. The Identity "of" the passage from one contrary to another, which remains itself without being alienated in this passage, or which is nonpositional (of) self, is not reflected in becoming and forces this becoming—or lets it be—to remain "pure" becoming, absolutely stretched out or flat, without horizon or loop, without cusp point or point of transmutation—a *static* becoming, a becoming *without passage*. Philosophy is the *looping* of time in space, for example, and vice versa; theoretical knowledge emancipates them "from each other." This means: from their reciprocal relation rather than one (from) the other.

Can we then speak of "becoming"? Or else, once the *philosophical* contraction of contraries, their syntax, is removed (their fold, coupling or doublet, their dialectic, etc.), must we speak of a contiguity of terms, a contiguity that excludes a priori every synthesis and every transcendent bond (dialectical or otherwise) that would claim to survey them? We have to speak once again of chaos; the sole organizing law, the only truly immanent economy of the terms individuated in this way is the fractal structure of Unilaterality, which alone determines them as the ingredients that form chaos. Contiguity is a last form of continuity; the juxtaposition of *relatively* indifferent terms is a last "logic." For chaos to exist, indifference must not be relative, reciprocal, "between" terms; it must come, after the One-Multiple of-the-last-instance, from the fractal structure that determines them (precisely as *terms*) by being indifferent to them. The variously articulated relations that compose the Multiplicities, the Differences, the Catastrophes, the Folds . . . are lifted here for the benefit of each term's Identity as 1. Even Transcendence's content is seen or received "in-One" and receives an Identity that holds "for" it without being dependent on it.

GF is not Identity that is subjected to logic, to its positional transcendence, and that has become *Principle of Identity*—or subjected to ontology. On the contrary, it is as it were the *logic of Identity*, the law that imposes this logic on thought.

On the other hand, it is tempting to speak of a fractal time, of an absolutely originary fractality of time. In reality, this generalized fractality holds for space-time; it is even their point of radical or of-the-last-instance identity. No doubt it is a matter of a time without ontological temporality or of a temporality without transcendent temporal things that add their own time to an indeterminate temporality-generality. Yet the same is true of space: a space without ontological spatiality or a spatiality without spatial things. Transcendence's most primitive essence, its phenomenal kernel, is the fractality we have described. It rests on a profoundly antiamphibological Identity, but it holds for the "contraries" that are space and time just as it holds for all the others. We thus eliminate the old hierarchies, antinomies, and aporias of space and time that philosophy produced in its history. This is why the Greek obsession with *becoming* and its *return* in modern and contemporary philosophy as the primacy of a temporalizing, originary, and open temporality of space itself must be now dismissed. We have described the phenomenal traits of theoretical representation: as nonhorizontal sprawl, as recto-without-verso, as nonlooped-opening, as unlimited becoming in the sense of *each-time-one-time*, etc. The categories of becoming, of passage, of continuity, of discontinuity, of temporalization, of temporal ekstasis . . . are philosophical categories charged with dominating Multiplicities, Catastrophes, and Singularities in the philosophical sense of these terms. And those categories are marked by a certain primacy of time over space. From this perspective, we will prefer more "fractal" descriptive categories like identity, unilaterality, spread(ness), flatness, "without" (horizontality-*without*-horizon, etc.). Fractality is "identically"—but in-the-last-instance—space (and) time. This prohibits every hierarchization that would be merely a way of reintroducing continuity where only the law of unilaterality reigns. The variously balanced amphibologies of time and of space are dissolved or suspended in the real theoretical labor. It is true that GF also consists in moving from a time *blended* with space to a purer time—as Henri Bergson and Martin Heidegger sought, although they did not succeed to the same extent. But if the "pure" or the "original" still forms a system with their mixture, on the basis of which it is extracted and which it conserves as a constraint that is impossible to absolutely lift, then it merely hierarchizes in another form what should no longer be hierarchized. Time and Space gain their *identity* as "terms" only if what is at stake is not their identification, but a "last-instance" that lets them be as such and protects them from their reciprocal determination, which would only inhibit one through the other. Here again, it is perhaps in terms of fractal chaos that the reciprocal "nonrelation" of a fractalized time (and) of a fractalized space must be thought every time, in a unique and uncommon way. This is fundamental for the description of the new fractality: the fractality of theoretical space-time and, in this way, that of language. This *fractality* is no longer manifested in the World as geometric fractality or in philosophy as temporal fractality. It is manifested as the purest form of transcendence itself; as the spatiotemporal opening, but without "ekstasis," without topological, catastrophist, revolutionary encystment. It is a matter of a *delooping* (in all the senses of the term), a deteleologization of time and space as well as of thought and language.

GENERALIZED CHAOS; PHILOSOPHICAL ORDER
AND DISORDER; FRACTAL ORDER AND CHAOS

The problem of GF's "intermediate" or nonintermediate character contains yet another aspect and has to be reexamined. GF is not only an original "intermediate" being, asymmetric between Identity and philosophy. This time it is intrinsically intermediate inasmuch as its essence is the strict Identity that remains (in) itself. How, by what effect, can an Identity remain intact, unalienated, at the heart of a coupling, of a milieu formed at the origin of two terms? Identity is manifested by its effect, Unilaterality as GF, but what is GF's internal concrete content?

The material on which Identity acts is what we called philosophy's mixture or doublet—its semifractality. Identity annuls or invalidates philosophy's spontaneous self-reflection: if there is a fractality "of" philosophy, it cannot be discerned unless philosophy stops thinking itself. But the residue that subsists is still the mixture, the doublet-form, though it is "fractalized." It thus acts positively, as we saw, on a coupling of opposites, which has simply become sterile. There is no reason to choose one of the two terms rather than the other. Identity does not choose, and GF is not a decision, an irregularity of this type; it is, rather, a finished stricture, an absolute constriction, in short an Identity. Each of the two terms is purified of the other, of its blending with the other, more exactly: of the mixed form itself. It is identical (to) self in-the-last-instance and without passing through the mediation of this form, without drawing itself on the horizon of this curve. It is a *term* in the full sense of the word, which has no need to be mirrored and alienated in the other because it obtains its being from Identity itself. It is not reduced to Identity, but enjoys the plenitude of its quality or its nature. It is a relatively autonomous *order, instance,* or *sphere.* The rule is to describe, without any blending, what each term of the fundamental couple becomes according to Identity's guiding thread, without nevertheless forgetting that the fractalized object is always composed of two terms and that each of them is therefore found in fractality and has to be described (for example, what is the "scientific experience" that is thought under the condition of Identity of-the-last-instance? And what does "theory" become under the same condition?).

It now suffices to combine the identical-in-the-last-instance terms and the structure of Unilaterality, of the GF that is related to them in the intentional, nonobjectivating mode and that functions as their fractal "law." GF is the reason that distributes philosophy's basic terms outside philosophy's ultimate law and now according to the law of the greatest irregularity. This is why the terms supplied by philosophy are henceforth independent of the law of the mixed form, of every *philosophizable* syntax (this law is not arbitrary, but applies to every syntax given with the World and Transcendence). However, this disorganization of the philosophical is not a bad nihilist chaos, because this manifold of determinations is distributed according to GF's now immanent structure, which is related identically to all these determinations. *We call chaos, in a more complete sense than the initial and always scientific sense, the fractal and immanent distribution of a manifold of determinations.* Here chaos is no longer only the One-Multiple but also Unilaterality insofar as it is exerted in toto within the manifold of philosophy. Philosophy, with its manifold of decisions, dimensions, and so forth, no longer intervenes as such in this case; it is *reduced.* It is indeed the philosophy-form that is GF's material, but the philosophy-form that is reduced to a manifold without philosophical relevance. *Chaos* is then GF's relation to its *internal* material, to the *reduced* manifold of scales, forms, and dimensions of the philosophical. Quite simply, it is "the whole" of science that is fractal, what we elsewhere called the Universe rather than the Cosmos. In philosophy, chaos is said of the *World*; in science, it is said of the *Uni-verse*.

There are evidently two concepts of chaos as of every thing:

1/ A chaos through autodissolution, or else through autoaffirmation of amphibologies, mixtures, or blendings—amphibologies that are constitutive of philosophy and that are simply proliferated, accelerated, and aggravated by philosophy itself as the factor of disorder. Hence the neighboring concepts or modes of this bad philosophical chaos: Differe(a)nces, Chaosmos, Multiplicities, Language Games, etc. This unfinished chaos corresponds to a limited destruction of the philosophical order, of its most representative and superficial forms, and to a reaffirmed respect, a reaffirmation of this order. *Philosophy produces neither a radical, absolute chaos anterior to it nor some strict, unilateral, or noncircular order, but a simple dis-order in view of a normalization of the real and science.* A semiorder, semichaos; a tendential yet limited dis-location, barely touched on, in which philosophy finds the reasons and means of its survival and its permanence as a tradition.

2/ A non-philosophical chaos whose cause-of-the-last-instance is the nondecisional Identity (of) self; whose specific element is the irreversible, absolutely fractal gap of Unilaterality; whose effect is to be the mode of existence, under its conditions of philosophical origin, of generalized fractality. It attests to the philosophically unintelligible or inadmissible fact that *the purest order, the simplest, is the germ of the most acute fractality and the root of chaos. The real chaos of science—not of knowledges in general, but of knowledges from the perspective of their theoretical relation to the real—is more than the opposite, it is the real critique of the philosophical dis-order.* This disorder corresponds to the fact that an order that is transcendent (and thus woven from disorder) is itself posited and presented as real and unintelligible by itself. Philosophical dis-order is merely one particular order, which is posited as universal and real and then lets itself be corrupted, dislocated, and dissolved by what it engenders, but certainly not destroyed. Science's chaos no longer forms a system with philosophy's bad nihilist or merely counternihilist disorder; instead, it forms its system with the immanent One-Multiple and the primitive order that fractality introduces.

So we will not conflate fractality with the aleatory, a passably negative and reactive concept that presupposes a rational norm. The fractal is and describes an essential, if not "normal," *state of affairs*; it integrates with science. It does not suppose as given or valid an order that it would perturb—in the manner of the aleatory. The fractal is itself the order: *chaos has its essence—but in-the-last-instance alone—in the most primitive order*. If it were not "in-the-last-instance alone," then thought would return to philosophical errancies.

The Identity of Irregularity, as the specific qualitative structure of scientific knowing, is thus chaos-without-logos, the fractalization of every chaology, of chao-logical Difference itself. Of course it is hardly close—since it is its real critique—to the philosophical *chaosmos*, the doublet of the old philosophical Cosmos and its contemporary, but always philosophical rejection. Let's call GF under its philosophical conditions of existence "generalized chaos." GF registers the destruction of the *autoposition of the Other* for the benefit of *Identity-of-the-last-instance-of-the-Other*, just as generalized chaos registers the destruction of the philosophical semichaos that exists only through the autoposition of the metaphysical autodissolution.

Measured against the irreversible fractality of scientific knowing, philosophy appears as an enterprise of normalization of science and, perhaps, of false liberation, false or grounded in an ignorance. The most creative irregularity belongs to science; it is not the irregularity of the philosophical decision. Science's chaos is ultradeconstructive. Thought's fluctuations, productions, and creations (for example, the new theoretical *discoveries* or *objects*) should not be absorbed. They are not themselves the exception; they are the rule or the essence. The "exception" to the rule is not an exception and is more than the rule: the cause that turns transcendent regulation into a normalization . . . The opposition between the *variable* and the *stationary* state, the fear of variation, of shock, of aggression, the rejection or the negotiation of novelty, which is then treated as margins, differences . . . all this is surmounted. The philosophy of Multiplicities, of Catastrophes and Differences, of Inconsistency, of Fuzziness amounts to under-/derealizing irregularity for the benefit of the philosophical form that remains the game's master . . . This game is still that of philosophy as the autoposition of Transcendence (of the Other in this case). In contrast, the power of the Other must be itself determined in-the-last-instance by Identity, the key that unlocks the kingdom of generalized fractality.

Science's mechanism is intrinsically chaotic, not only infinitely open-without-teleology, but fractal in-the-last-instance, i.e., through and through. Scientific knowledge can fluctuate, vary, and be renewed for apparently *external* reasons; they stem from the emergence of new phenomena (experiences and theories) that are from the start virtually philosophizable. And science has always struggled against this, whether expressly or not. But these reasons become internal to the suspension of their philosophical sense; this suspension constitutes them as the *data* of an *experience* in the scientific sense of this term. No external—for example, techno-political—stimulus can enter into competition (with respect to science's essence) with its fractality, which is identical (to) self and fractalizes these stimuli in their turn. This advent of scientific fractality takes place without *returning* after each variation—or each new knowledge validated by any means whatever—to a *stationary* or *normal* state. The paradigm of *crisis* and of *normal science*, like falsifiability, and in general the metaphysical constructions of Karl Popper, Thomas Kuhn, and so forth, are unitary, transcendent, and foreign to science's real labor. If science is a chaotic process, even in theory, then all the transcendent constructions of epistemologies are useless and uncertain—if not as normalizations of science.

We have summarily described a semifractal economy and a semifractal anatomy of philosophy. Invariant forms in rare numbers regularly emerge from the philosophical Tradition: the circle, the abyss, the fold, the doublet, the (double) band, the arborescent network, or the rhizome . . . But they are hardly fractal and not quite unilateral or irreversible, though they are already transcendental rather than natural. But here again they are seminatural, semitranscendental. Philosophy is an idealized double of nature and entirely lacks spirituality and interiority on one hand and scientific immanence on the other. The functions of philosophy's semifractal lines and surfaces involve the capture of a maximum of information, resistance to external disturbances, struggle in general, and concentration of information (fold, double, continuum, etc.) more than they involve order, distance, economy, Occam's razor, and the destruction of every teleology—these are science's functions. Philosophical fractality partially imitates nature's. It is ordered to teleologies; it proceeds by redundancy as much as by strict or "identical" self-similarity, by Sameness rather than by Identity; by tautology and accumulation of the reserve of knowing and of culture rather than by poverty, order, and simplicity.

THE GENERALIZED CHAOTIC DISTRIBUTIONS OR NON-PHILOSOPHY

Perfect yet artificial fractal curves, constructed through recursive procedures of infinite interpolation, through the introduction of new ever smaller irregularities and thus conserving to infinity the property of homothety, can exist in geometry. Yet they only exist in a very improbable way in nature where the scale changes are punctuated by changes in order, reign, or domain, with the same phenomenon moving from the macroscopic to the microscopic, from the meteorological to the chemical, from the molecular to the quantum . . . The "passage to infinity" is prohibited in nature. This perhaps helps explain the usual presentation of fractality in the form of *clusters* or regroupings of the "same" fractal object, which are in general strongly lacunar but also regularly dispersed.

From the perspective not only of its support but of its principle as well, fractality presupposes some intermediate or transition zones and would annul itself if, for example, an irregular line really and adequately filled a whole area and became continuous again in its own way. Natural objects combine several fractal ensembles ("hierarchized clusters"); this prohibits or renders difficult the fixing of a scale on which an adequation for a unique object would take place. We cannot compare in a direct, intuitive way a line's length and the surface it partially occupies or fills. There is a phenomenon of "clustering," of grouping of objects in distinct clusters, which appear as the change in the scale of magnitude. It seems that the same must be true for a GF where the "non-philosophical" statements thus produced are not only distinct but must form fractal aggregates, cluster-identities. Regardless of what happens between geometry and physics, our problem is to describe what happens in thought between science and philosophy and how the GF is distributed.

On one hand, the object "philosophy" also knows scale-changes (we can examine or survey it while participating in it—"meta"—with different degrees of resolution) and successive appearances that are increasingly complete or increasingly fuzzy and incomplete, dimensions that constitute its space (a theme, then the Dyad in which it is included, then the philosophical Decision that structures it, etc.). Philosophy also knows— provided we take it as a global reign, as a tradition—changes in decision, a "same" object, "critique" for example, going from the Kantian to the Nietzschean critique. . . . *From the perspective of its materials*, it does not seem possible that GF would lead to an infinite unhindered dispersion and seems that it too has to be distributed or grouped in more or less hierarchized lacunar clusters whose last reason would be this specific heterogeneity of philosophy. A reason that is different from fractal irregularity and that limits it in some way.

On the other hand, GF exists in the form of *terms that are strictly identical each time* in their specific qualitative, autonomous nature; they are clustered in strictly nonhierarchized forms by the absolutely dispersive a priori fractal structure. These determinations no longer float in a universal element; they are, so to speak, the manifold of the Other itself, the one it aims for in an in-objective way. Not an identity-*of*-the-Other, a Same, but an absolutely and thoroughly fractal Other, which is only identical in the last-instance and exists in the form of this radical chaos. Along with Identity, fractality is chaos's determinant reason, and chaos is fractality's concrete existence. Inside *chaos*, which is the chaos "of" philosophical determinations, philosophy stops reigning. The continuous topological relations of connection or of vicinity, the singularities, the catastrophes, the games, the partial objects and flows, the differences and differends are invalidated. The old terms of philosophy's representational content are no longer linked or connected, topologically or otherwise, by some relation of co-belonging that would be proper and coextensive to them. They have in common only the fact that they are subjected to an identical fractality, which emancipates them from philosophy's continuous curve. Fractality alone, the most simple or the most irreducible to every philosophical curve, distributes these determinations. Just as GF eliminated for its part the geometric transcendent and symmetric figures of fractality, so the radically thought chaos excludes for its part the distribution into clusters and lacunae.

More exactly, and in order to take this double constraint into account, the clusters, lacunae, and hierarchies, which can always be philosophically regrasped, are now only *terms* or *determinations* distributed by a chaos so fractal that it suspends their operative character. And so we will not imagine— this would be one more philosophical imagination—that thought has an interest in adding the clusters and the lacunae, the promontories and the bays to the already old philosophical imagery of holes, sheetings, Möbius strips, valleys and rivers, ruins . . . and to the even older imagery of circles and vortices. It is a matter of *thinking chaos* and of ceasing to imagine it, for generalized fractal chaos is more complex than the chiliagon and defies philosophical understanding itself. Science's "dimension" is traced by the axis that goes from Identity-of-the-last-instance to its correlate of chaos. Chaos is what we see "of" philosophy when we situate ourselves in this wholly "internal" fractality.

CHAOS AND THE CRITIQUE OF PHILOSOPHIES AS "THEORIES OF KNOWLEDGE"

If scientific representation is essentially chaotic (and not chaological), then it is the real critique not only of philosophies that are, above all, "theories of knowledge," but of all philosophies insofar as they always, and in any case, have a certain conception of knowledge and of science.

Fractality does not only change its nature when it passes from the geometric and/or philosophical Same or Whole to Identity-of-the-last-instance. It also changes its field of exercise or its object: it passes from natural objects to thought objects, but above all from objects to their representations. From this point of view, we can detect some hesitation in fractality's geometric or restrained theory: fractal is said at times of the natural object to be modeled, but at others also of its mathematical representation, which is deemed to be more adequate in this case. From there to the conclusion that the object itself is fractal, one only has to take the philosophical plunge of circularity in the mirror of the object and of its representation. For instance, we may call "fractal object" any "natural object that can be reasonably and usefully represented in a mathematical way by a fractal set" (Mandelbrot). As always, this term *representation* receives a very "representative" or specular use. It is conflated with the supposition that fractality is ultimately a form "common" to the object and its representation, which is its more or less wrought "tableau." This is the heart of the theoretical *intuitionism* of geometric fractality. It cannot be up to us to decide in the present case whether fractality is the fractality of the object or of its representation, even though the conception of science at work is here quite clearly that of a transcendent realism of perception and even though science is understood as a more or less specular double of its object. This approach to science is possible only if the object and its properties are in some sense reified and posited as objects "in themselves," as common sense wants to do. Theory is then no longer intrinsically and qualitatively distinct from its object.

By contrast, the theory of science we are defending is a realism in-the-last-instance alone and not a theory of perception or of transcendence. So much so that theory shares no *common* form with its object—which it does not modify. So much so that this object's presumed properties "in themselves" are only older and more elementary knowledges that are reified and realized. The presumed object "in itself" must be dissolved in the real-of-the-last-instance and in objective knowledges. Science is the thought that remobilizes these presumed properties of the object as knowledges and makes them serve as materials for other, more universal knowledges. If these two states of knowledge share a "common form," it is the form of theory or of scientific knowledge, which is no longer specular, bilateral, and transcendent, but unilateral. Instead of a third and transcendent form common to the object and its representation, effecting their synthesis for its own benefit, there is the intentional-nonobjectivating relation of theory—a kind of "common fractal form," but absolutely immanent and not transcendent to these two knowledges, a relation that individuates them as terms, as knowledge-identity, at the same time that it fractalizes them. This amounts to saying that instead of this specularity, this third-mirror in which the object and its representation communicate and are exchanged, there is a chaotic or fractal dissemination *of the "object" and of "knowledge," i.e., of thoroughly universal knowledges*. In this way, generalized fractal chaos destroys, to its root, the possibility of philosophies of science and not only explicit "theories of knowledge." As such, we will say that, in becoming universal or generalized, fractality has passed from the object—and its intuitive images—to its representation alone. But this representation should be understood as an autonomous sphere or an identity of theory, absolutely foreign—precisely because it, and it alone, is intrinsically fractal—to its objects as *data*, to what serves it as materials. This fractality of theory, which is its "specific difference"—and more than its difference—from its object as it is presented from the start, is what fractalizes all the objects-of-knowledge and plugs them back into science's living circuit: even these old reified knowledges, the *natural* or *perceptive* properties of objects—including the sensible or intuitive fractality of Britain's coasts, of clouds, and of the Ocean.

Furthermore, a clear solution is provided to the similitude—from now on, the immanent fractal identity—of the object and of its representation. The geometrician's hesitation on this point marked the problem in a symptomatic way: there must be something like a property of resemblance, at least the equivalent of a similitude between the object and knowledge. But instead of reifying this resemblance in the form of the mirror-agent presupposed by all philosophies, Identity-of-the-last-instance alone is enough to explain that fractality can be said (both) of the object (and) of its representation, of the thing and of knowledge, though it is only the fractality of knowledge. This is why by "GF" we mean as much the property of an object as the property of the knowing of this object. On one hand, the object is itself only an old fractal knowing whose fractality was effaced or denied. On the other, two knowledges are fractal not through a common fractal form that divides up two supposed givens between them while remaining itself one or undivided, but through the investment of a one-or-undivided-fractality in a material that it and it alone *individuates* or makes visible, not so much as 2 but as 1, 1, 1, etc. Chaos is not confusion. On the contrary, it is what individuates by means of fractality itself philosophy's confused blendings or amphibologies—its "order."

If philosophy is a statistically regular curve, despite its angularity—but it is always bilateral rather than unilateral—we will not say, on the other hand, that science too is a curve, that it is simply more interspersed with angles or points: the theoretical discoveries. In reality, science is a chaos in which even the local curves are fractal events or absolutely irreversible catastrophes. It is impossible to smooth out science in a becoming or a teleology as philosophy does (these attempts, nevertheless, are not lacking). Science does not have the nature of a *reserve* as philosophy is; it has the nature of a *chaos*.

The philosophical Tradition can be analyzed in an infinitesimal way. It subsists and resists and continues to accompany, with its more or less breached totality, the details of its decisions. At every point of its decisions' curve, it is possible to extend these decisions with a straight line (a new philosophy) or a linear development. So much so that the Tradition is sprinkled with these altogether straight doctrinal lines that approach Tradition in each of its decisions and give to philosophy a simultaneously pointed and continuous character. We know, moreover, that the philosopher contrives to trace two lines from the same point: the one circular, the other straight (Plato, Leibniz, Heidegger, Nietzsche, Deleuze, Althusser, etc.).

Yet science is so profoundly irregular *in its very essence* that it is detached from the start from philosophical idealizations and makes use of idealization, of regularization, only as a local procedure. It is incapable of outlining a coherent, teleologically dominated future for its work—if not by entrusting itself to philosophy and the State, united in a prestigious and intimidating alliance. The angularity of knowledge in its relation to the real and, *thereby*, in its relation to another knowledge is one-sided, so to speak, and traces only irreversible paths. This does not mean that it is impossible to survey them, to recross them in reverse or to "philosophize" them. Science is not the development of *possibilities* to be realized or of a virtual to be actualized. It is ultimately—to use an ontic or regional metaphor—a Brownian, yet transcendental movement, a movement that concerns knowledges themselves in their universality while existing only under philosophical conditions of existence. So long as the description remains faithful to the essence of science and does not found itself on some objective knowing, which is philosophically objectivated a second time or reposited, fetishized, or factualized in the form of transcendent *fakta*, it can plunge into the details of knowledges without ever apperceiving these phenomena of totality, of systematicity, and of continuity that philosophy believes it detects in these details. The description does not follow any line and, as a result, a line that weakens its heterogeneity or its multiplicity. Each "term," each knowledge, obtains its sufficiency from the Identity-of-the-last-instance, from the cause of science rather than from an ontological relation to others.

Whereas philosophy is thought-through-system as well as through the proliferation and alleviation of systems, science is thought-through-chaos. What distinguishes them is by no means the artificial couple thought/knowledge, but two heterogeneous types of essence or of cause of thought-knowledge. Scientific practice reveals a granular—yet transcendental rather than physical—structure of knowing; a subatomic identity that is no longer homogeneous with the atom, but is "utterly" distinguished from every ontic or ontological-transcendent identity. It is an *identity-of-the-last-instance*, the ingredient of chaos. It is enough to think a difference without a line of differentiation that accompanies it and to reserve the right to an analysis and a synthesis external to the very event of knowledge; without a possibilizing foundation that immerses this event in the possible. The continuum, but also difference, the inconsistent multiple, language games . . . are artifacts, crystallizations, and smoothings of certain phenomena that are too accentuated for the philosopher to overlook them; the operation of an arbitrary and fascinated vigilance, obsessed with life and especially with survival.

Left to its autointerpretation, philosophy envelops itself with a kind of unlimited skin (surface, plane, plateau, slippery or rugged ground, etc.), through which it simply slides on science's "fractal" asperities and shelters itself with them. Granular skin, differentiated into organs and smooth zones of reception—but it is a skin, the interface with another philosophy; a skin that philosophy does not live so much from within as surveys, anticipates, and projects. Science is another experience of the real: not through *interfacing*, but through unifacing, as we said, through a fractality—the Other itself—that is directly felt on the philosophical skin.

GF'S FIELD OF APPLICATION AND PERTINENCE

We elaborated GF's concept by making use of the "categories" of Identity and Difference and of categories affiliated with Difference (multiplicities, differends, différance, language games, etc.). They are types of "meta"-philosophical transcendentals that have a reflexive vocation and that allow us to clarify and thematize amphibologies in general (not only those of the understanding, the Kantian amphibologies). As a result, we can better understand that fractality itself, as we described it, participates in this metaphilosophical—now: *metascientific*—function or serves to describe knowing in its essence as well as to critique the great—philosophical—amphibology of science and of philosophy. It belongs to the theory of science and thereby to the science of philosophy. Without a doubt these universal fractals can only be the fractals of a knowing—it is now the object "theory," rather than Britain's coast, that *is* intrinsically fractal. It is a matter, as we said, of a theoretical mutation that touches on science's essence, that saves it from its epistemological capture and in this way reaches philosophy itself.

Yet if GFs are objects of knowing rather than of "nature," this does not mean that they are obtained through the autoreflection or autoposition of geometric knowledges, in the form of a *faktum*. They have no doubt the *essential* properties of self-knowing—they pertain to the theory of science—and, being related to knowing rather than to the object, they are transcendental. This is a generality that can also receive a philosophical sense, that can be understood precisely as a process of autoreflection, but it is in fact a matter of an unreflected knowing, nondecisional and nonpositional (of) self.

Identity's fractals are open only to a pure description; they are immanent phenomenal givens that no philosophical decision of generalization affects. They are from the outset the "most" universal. For universality is not in this case a question of degree, but of a "qualitative" definition or of a definition according to the essence: it is not divided into generality/totality; it is undivided in-the-last-instance as a unilaterally open sphere of objectivity. We see each thing "in"-the-last-instance without each thing having the structure of a reflection (even an objective reflection); without each thing becoming productive or participating. It is a contemplation that only the static phenomenal givens can fill. Even "becoming," time, and practice are pervaded by this static phenomenality that corresponds to their fractal structure. We would readily speak of a fractal vision or mysticism if these terms did not convey the worst philosophical confusions.

What does it mean to say that the fractal theory is generalized? We can always imagine a "superior" fractality, which is specific to Being or philosophy, obtained through idealization and interiorization of geometric fractals, "superior" insofar as it is a mixture of geometry and philosophy. But we showed that this concept, which subtends the contemporary philosophies of singularities, is in fact a semi-, at once an under- and an overfractality in which fractality is impoverished and effaced (qualitatively; it is not a question of degree). It is more rigorous and fertile to identify a full or finished fractality, absolutely universal de jure, a fractality that can *apply to* philosophy itself from the start rather than remain caught under philosophy's law and thus immersed in an ever indeterminate generality.

It does not result from a supplementary, philosophical idealization of geometric idealities; this idealization would, paradoxically, turn these idealities into *faktums*, into supposedly autonomous *rational facts* that can be extended by a philosophical reflection. For instance, we have treated the geometric property of *internal homothety* less as a natural property (which would only be a *faktum* produced by and for a philosophical decision) than as a simple scientific knowledge, a local theoretical tool. And so we considered it as the simple *indication* or *material* of an essentially theoretical problem to be resolved (rather than as a rational *faktum*): the problem of the fractal constancy of science itself. We have thus avoided extending it in the (ontico-ontological) mixture of a *transcendental homothety*. Its equivalent for science is Identity-of-the-last-instance; it is no longer a homothety, which would presuppose operations of displacement and inversion. We have ceased to capture science philosophically, and have began treating it as an unconstitutable phenomenal state of affairs that can only be described. By the same token, we have not interiorized fractality proper, the condition of irregularity, to the concept; we have not divided and sublimated it with a philosophical decision and thus produced a universal fractal *operation*. But we have treated it as an indivisible body or drive, as an *undivided distance* without reverting or returning in any way to Identity. Everything is here lived or received by the cause (of) science in its unalienable mode—this is the sense of "in-the-last-instance"—so much so that everything in science is a purely immanent = indivisible phenomenal state of affairs, even when, like theory, it proves to be complex.

The concepts of generalized fractal and of chaos are more powerful theoretical (as well as artistic) tools than "differences," "inconsistent multiplicities," "games," "turns," "disseminations" . . . because they are theoretical rather than philosophical and because they define a fractality in relation to philosophy itself, considered globally. On the basis of Identity-of-the-last-instance's incommensurability to Being, it is the whole logic and ontology of mathematical multiplicities, the topology of differences, the philosophy of fractals and the philosophy of catastrophes—not their mathematical or geometric bedrock, of course—that appear as half-solutions and do not do justice to the scientific sense of purely geometric or set-theoretical givens.

In a correlative way, GF sees its relevance limited to philosophy, i.e., to a discipline that uses natural language, is not reduced to this language, and can even determine it. It is strictly "transcendental" in the sense of the flawless immanence of its cause. It is thus not empirically identifiable or locatable in the World, in History, in Power, in Sexuality, or in Language, because it is fractality-in-Transcendence itself. To fractalize every form of ontological transcendence, *to fractalize Being itself rather than to defer or reserve it*—this is GF's effect. One particular mathematical or physical theory cannot therefore exemplify it. On the other hand, it is fractality *for* Transcendence, for the World, and, in particular, if it cannot be found within language, which does not provide it, it is "destined" to shape natural language, at least insofar as it is required and included in philosophy and ordered in the sense that a philosophical decision can be superimposed on it. It is not an ontic science. It is pertinent for natural, physical, or mathematical, but also social or unconscious phenomena only to the extent that they are prima facie given as clothed in *sense*, coated with the philosophy-form. More rigorously, it is not a matter of a regional scientific fractality that is applied to philosophy, but of the fractality of "science itself," of theory as such, thus of *its scientific sense*. It is on this "transcendental" condition that it can *apply-to* philosophy. "Science itself" is not "fractalized" or represented by a fractal model; it *is fractal* in its very essence. On the other hand, it is philosophy that, not being spontaneously fractal or being so only in half, is *fractalized*. It is fractal theory alone that thus possesses an original and de jure critical dimension in its relation to the manifold or to the object, precisely because—unlike its geometric and philosophical forms—it is founded on the distinction between orders or spheres of reality.

GF's status and its type of pertinence have aesthetic consequences. If GF has to have a more direct, particular affinity with certain phenomena, these phenomena will be literature and poetry more than painting. By right, all arts are equal before it, since everything must pass through philosophy's mediation (they are virtually philosophizable). In fact, GF's circuit of access to phenomena is evidently shorter when it is a matter of the arts or of disciplines that mainly use natural language (law, ethics, etc.).

But in all respects GF represents a *theoretical tool* (theoretical and not technological). It has at its disposal the same type of power as the geometric form: a theoretical and not philosophical power. Far from remaining narcissistically in itself and circularly *auto*applying to itself in an indefinite, repetitive, and sterile way, it enjoys de jure a theoretical intentionality, an intentionality of knowledge. It has an object distinct from itself: philosophy. It enables the theoretical knowledge of philosophy, its transformation into a *scientific continent*. It produces new, emergent statements irreducible to philosophy's *data*, to their spontaneous "ideological" representation (philosophical faith, the Principle of Sufficient Philosophy). It is thus not a matter of a simple transfer, like a technology-transfer, with the problems of inadequation that results from it, but of a theoretical mutation within the concept of fractality, on one hand and, on the other hand, of the recognition of a specific intentional aim directed at the object "philosophy," an intention of knowledge that belongs de jure to this transcendental concept. We will not conflate this theoretical elaboration, this *rupture* in the theory of fractality, with the conceptions—stemming from the epistemological disaster—that reduce scientific labor to an activity of importing-exporting of concepts. We can invent or at least discover philosophy as a scientific continent on the condition that we abandon not only these practices, but the very ideal of philosophy-as-rigorous-science for the wholly other Idea of a science-of-philosophy, which requires a recasting of the theoretical tool: with the double goal of giving it a truly scientific and not philosophical pertinence and of rendering it adequate to its object, which is not arbitrary—no more here than elsewhere. GF's concept, which condenses our previous research in the manner of a concentrate of non-philosophy, responds to this double objective.

GF's concept has made us realize in hindsight that what we called *non-philosophy* was already a fractal *type* of practice of philosophy. The practice presupposed a remodeling of this concept, which is thus "spontaneously" extended to natural language and to philosophy. The produced statements (and, as a result, also those that form its theory) no longer respond to philosophical logic, to the norm of its statements' production and admissibility, because they are produced as reproducing a structure characterized by the internal Identity of its inequality, despite multiple decision-and-scale variations. On the other hand, one of the major interests of this theory, alongside its immanent "application" to natural language (rather than to space and to ontic objects), is more clearly to mark the difference in nature between philosophy and science. As the criterion of space or of knowing, it distinguishes this space absolutely from the philosophical space, which is *not fractal but coordinated or mixed*.

PART III

PRINCIPLES OF AN ARTIFICIAL PHILOSOPHY

5

UNIFIED THEORY OF THOUGHT

TECHNOCOMPUTING REASON
AND THE NEW ORDER OF INTELLIGENCE

TWO reasons lead us to revive the classical problem of a science of thought and force us to abandon the solution philosophy traditionally invokes with the names of "logic," "science of logic," or "doctrine of science" under which it presents itself in person.

The first is superficial and situational: the irrefutable emergence and extension of a technocomputing experience of thought in the form of extra administrative and "intelligent" uses of informatics. In what capacity can the rise of the information technologies of intelligence and reason ("artificial intelligence," "cognitivism," and their future relays) still be interpreted by philosophy and dominated by its authority and procedures? Philosophy always "reduced" the autonomy of mathematics and arithmetic, casting them into an inessential "phenomenality." But the inverse reduction, that of thought to reasoning and of reasoning to arithmetic, which receives unheard-of technological means, forces us to reconsider philosophy's relation to science and even to technologies. The technocomputing drive has perhaps an origin that its "metaphysical" sense does not exhaust.

The second reason is more fundamental and indirectly determines the first. It is the problem of scientific thought alone and of its relation to philosophical authorities. If science, as we are trying to show, is an autonomous experience of thought, autonomous vis-à-vis philosophy, if there is a genuine thought of science, i.e., a specific relation to the real, determined by itself, that knows itself to be so determined without passing through philosophy, then all the old divisions have to be reconsidered and the general economy of the field of thought has to be disrupted.

Of these two reasons, the second is essential or "transcendental," i.e., determinant of the degree and forms of thought's *reality*. The first is only "occasional" or "empirical," if this term still retains some sense here. The appearance of the technological pole between the philosophical pole and the scientific pole is only a symptom and is not enough to seriously endanger the traditional philosophical authority and its legislation over thought. The emergence of technocomputing reason is an indication, one that something else, science rather than technology, seeks to make visible in its specificity against the philosophical order that is regularly imposed on it. We have to do justice to what is not a sufficient reason. As such, before even being analyzed, the symptom maintains a precise discourse: the domination of Intelligence in our history. The age we are interminably entering is no longer the century of Critique or the century of History. No doubt we have no sure criterion for deciding that it is the century of Language and Communication or else the century of Intelligence. In fact, it is enough to allow the symptom to manifest itself and to form a situation. Subject to this condition, nothing prevents us from letting ourselves be affected by the technocomputing phenomenon as more fundamental than the phenomenon of Language and Communication. It seems to us that it challenges to the limit the domination of philosophy through the immediate bedrock and resources it discovers in science, whereas Language and Communication continue to move inside the sphere of philosophy, the sphere of its most invariant philosophical schema and its authority. This accounts perhaps for our impression nowadays that Language and Communication have not kept their promise and that they are exhausted as objects of science. A more or less rigorously founded science of history (Marxism) may have indeed existed, but no unified theory of the phenomena of language ever has, as we might want, unless we assume that this science of language is realized in the form of a general theory of communication.

Nevertheless, if we are entering a "century," the objective appearances suggest that it is the century of Intelligence. From now on, Intelligence arrives on the scene with the claim—grounded or not, this is perhaps not the essential problem here—of becoming something like the *principal* "productive force." A latent cognitivism, carried and accelerated by the technological drive, now traverses the philosophical and scientific as much as social and economic practices. Nothing prohibits our being affected by the amplitude of this emergence, by its dominant and "overlying" character as we were by those of History in the nineteenth century, since Intelligence announces itself, beyond Language, with the same force as History. Without doubt, this is a *dominant* rather than *determinant* phenomenon of the becoming-*real* of thought—the precise theoretical distinction between these concepts was invoked several times earlier. And if a rigorous science of Intelligence must at last be elaborated, Intelligence itself cannot be the foundation of this science but only its occasion. It is thus not in and around Intelligence that a general science or a unified theory of thought has to be constituted and, above all, founded. In any case, this project could have been carried out without it. On the other hand, as dominant phenomena that form a situation, the problems of Intelligence and of Thought are henceforth more fundamental than those of History and of Language. It is in this special capacity, at once crucial and secondary, that the new order of Intelligence (whatever the duration of its domination) serves here as our guiding thread for the constitution of thought into a "scientific continent."

OF SCIENCE AS THOUGHT'S REAL BASIS

How can we pose the problem of thought scientifically? How can we make visible—it is the same thing—the technocomputing symptom as nothing more than a symptom? Traditionally, the relation to science, to technology, and to philosophy takes place on the basis of philosophy and under its authority. That philosophy is posited as the real basis of itself is what we call and describe as Principle of Sufficient Philosophy (PSP). The project of a science of thought (including of philosophy) is the radical critique of this claim. It assumes that science is recognized as the only authentic real basis of our relation to philosophy and to technology. The problem is not simply to know whether we have in science the experience of the most real thought, of the thought that can serve as such a basis—the autodescription of science has supplied the proof of this. It is first and foremost the problem of what we might mean by this experience.

A thought is generally called "real" or has a transcendental consistency if it is, by its very essence, a relation to the real and to self-knowing as well as to the knowing of this relation to the real. But this is not enough to distinguish science (as authentically real thought) from philosophy, which also raises this claim, but reserves it exclusively for itself. Philosophy's peculiarity is precisely the PSP, the belief that it and it alone is a relation to the unique real and a knowing of this relation. A restrictive condition must thus be added in order to distinguish in a definite way the scientific form of thought from its philosophical form. Negatively, it is the exclusion of objectivating or autopositional transcendence—which is peculiar to every philosophy—outside the essence of science and even outside its practice, in which transcendence takes a nonobjectivating form. Whereas philosophy is related to the empirical real and to an ideal real (*divided unity of the "real"*) toward which it simultaneously transcends, scientific thought is already real by itself without having to pass through a transcendent reference in this autopositional form. Every science is the science of something "empirical." But it distinguishes between a given or an occasion—that is necessary as support and as material, but is contingent in relation to the real essence of thought, which resides in immanence alone—and the *real object* or *the scientific objectivity*, which is not reduced to a contingent given, which includes this given without finding its own essence within it. The object's *reality* is not codetermined by these empirical givens as is always circularly the case in philosophy. The description of science has clarified these points and dispelled its confusion with philosophy.

Positively, it is the very structure of the cause or the absolute "foundation" of science that excludes philosophical transcendence as useless or at any rate as not present. Whereas the philosophical real is unique-and-divided or is a *relation-to* the real, the real of scientific thought is *One* or *identical (to) self*. But it is not *also* divided *and is never in its essence* even an "internal" *relation* or a relationship. There is a point in science at which all the relations are "related" (but without any relation) to an experience (of) self that is not separated (from) thought. This self-enjoying experience of indivision is called nonthetic (of) self and forms the irreducible *reality*, the *real kernel* of thought. We equally call vision-in-One the One that would not necessarily be also accompanied by a Dyad or blended with it as is always the case in philosophy and suffices because it is nonthetic knowing (of) self. This real, which is not relational, which is simply *a-rational*, is the element of science as thought, as the nonpositional phenomenolization (of) the real, and not simply as the production of knowledges or assemblage of models.

Fundamentally obscure yet perfectly consistent thought, it does not pass through a representation of the real in order to become real in its turn . . . This new criterion of the real and of the representation that is cast outside the real allow us to include the philosophical and cognitive thought in this representation. Extending what philosophy calls representation, and of which it gives a very narrow definition, we will henceforth use this term to name every experience of thought (including philosophy) that is founded on or primarily uses transcendence in its relation to the real. On the other hand, given its ontologically decisionless and positionless nature, this experience (of) indivision, the identity of thought and (of) the real can be called *cause-of-the-last-instance (of) thought*. The nondecisional cause (of) thought does not fall under representation in general and in its vaster sense, since it is *index sui et repraesentationis*.

A science of thought-through-representation (including philosophy) cannot be founded in its turn, viciously, on a representative experience; it has to be founded in what is most *real before* and *in* representation. We also had to establish the project of such a science on its real basis (which is the nonthetic cause (of) thinking) rather than on thought itself (which includes a necessary and constitutive moment of representation). The problem is no longer to "possibilize" thought but to "realize" it—this is still a poor formula. Philosophy produces and consumes the artifact of thought that is Representation as autoposition, and the Human Sciences produce and consume this other artifact that is "Cognition." Yet the cause (of) thought is the sole content of phenomenality or reality that is indivisible and inalienable. It alone can thus serve as the real basis for our relation to philosophy and to cognition. *The experience of thought within the limits of science: this is the foundation of a science of thought that, given its absolute "universality" and priority, can include not only logos but also the technological and artificial forms of intelligence in its "phenomena" or its "objects."*

At the same time that the cause-One of science is the basis of our relation to philosophy and to cognition, the essence of science constitutes the *real object* that "first science" (as science of thought) describes in the first place. In other words, the very first task in a description of the essence of science can only elucidate in the One the *nonthetic cause (of) thought*, of every science, even if it is "empirical." This description is in its turn another science, but a science (of) the essence of science or "transcendental science." It is a "particular" discipline, located alongside others and free of any philosophical claims, but its object is what is more real than every Representation or Cognition.

THE PREMATURE OR ILLUSORY ATTEMPTS
AT A SCIENCE OF THOUGHT

If philosophy already treated the problem of the science of thought and the thought of science as "first" or as onto-logical, why should it be posed once again? If a science of thought that comprises a thought of science is philosophy in its entirety, how can we avoid adding a new variation to already elaborated solutions?

Not only is philosophy now presented simply as a particular solution to this problem: a *unitary* solution that, for this very reason, only covers certain phenomena, a certain experience, but not all the actual heterogeneous forms of thought. But we are also searching for a unified theory of these forms— unified rather than unitary (we will elaborate the difference between the unified and the unitary further on), unified on the basis of science alone. To be sure, we admit the ontologico-meta-physical solution; it exists, invariant and multiple. But to this solution we "oppose" experiences of thought, those of sciences and of rational "arithmetic," that this solution can only understand by reducing. How do we know that? The only positive reason for which we proceed in this way is science itself, which is capable of describing itself in its essence and of abandoning philosophy's ancestral authority. We can claim to pose this problem non-philosophically only if we have the means of showing that an experience of thought that was never ontological or philosophical is at issue in science. The project is then completely renewed, for the name *science* can no longer signify for us what it has always signified for philosophy: either the *empirical* concept of sciences or the transcendental or rather *metaphysical* concept of "absolute" science or, better yet, the hierarchy of these two concepts. We have to grasp identity before its division: the Identity of science instead of its "being" . . .

It is thus science, and science alone, that can denounce the previous attempts at a theory of thought as premature or illusory. They have never ceased and are extended toward the two competing parts that currently divide up the project. On one hand, philosophy in very varied forms (theories of knowledge, logic and science of logic, transcendental and empirical epistemologies, "experience of thought," etc.). On the other hand, "Artificial Intelligence" and "Cognitive Sciences," as the advanced tip of Human Sciences, which are presented as sciences of "general intelligence," i.e., of the real content of "reason." In this fierce competition the claimants share the fundamental presuppositions: those of "transcendence." That these knowings are simply "empirical" or more essentially "ontological" is not a pertinent difference for science, i.e., for an experience of thought that discovers its cause directly in a radical or "of-the-last-instance" identity.

What does a rigorous science of thought allow us to decipher in the present situation?

1/ From this new point of view and from it alone, of course, this science of thought is a project that has some effective forms. But it has not yet discovered its scientific form, the one that would be *grounded in rigor and in reality*. Above all, it has only had "ideological" external forms, manipulative and technological in spirit. Thought has to become a scientific continent. But this continent has only been occupied up till now by philosophy and some of its subproducts. Just as history was the object of ideological and appropriative theories in the form of "philosophies of history," which took the place of an absent science, so thought—but we are far from having punctured this subjection—is the object of philosophies and of technologies disguised as sciences, which manipulate it more than they contemplate it and occupy the vacant place of a science. We do not yet have the rigorous *concept* of the reality of thought; under this name we assemble representations, self-serving substitutes, unreal artifacts, and manipulations. With its scientific claim, cognitivism sounds a serious warning to philosophy and a call to finally think rigorously this very old and very new object. We confine ourselves to knowing thought in particular through its exercise, its practical manifestation, its works (philosophy and sciences as well as common sense). But these spontaneous practices are not equivalent to a science. Science's philosophical thought is not yet a science of thought. But if this concept is valid for sciences, will we say that it is not perhaps valid for philosophy, which is a self-knowing and a reflection even in its most spontaneous exercise? The exercise of thought in philosophy would have the privilege—an ontico-ontological privilege—of being the only one that knows itself, that is at the same time a knowledge of thought. Hence, for example, philosophical logic as a science of logic in the Hegelian style . . . But we want to suggest a whole other thing. In its essence, philosophy is nothing more than a spontaneous practice (a decision, a will, a scission—a practical moment through which it claims to determine and transform the real, even if it is *also* the theory of this essential practice). From our point of view, it is enough that "theoretical" thought be at the same time also a practice and especially, by its very essence, a decision, in order not to be a science, but a spontaneous or practical thought, with theoretical *aspects* or *ends*. Up to now this science has thus remained impregnated either with philosophical teleology and practical spirit or with manipulative and technological spirit—always the same self-serving obsession. One always wanted—though not always for avowable ends—to transform thought's reality, at the same time that one sought to know it—a pious wish that had to remain an infinite wish, an infinite desire . . . *A rigorous science is dis-interested, contemplative, and descriptive.* For reasons of internal rigor and of reality, it is necessary to absolutely exclude every teleology, every interest, and every will from science—conjugating on this point Spinoza, Husserl, and Heidegger. For scientific reasons, thought's self-knowing in the philosophical mode cannot be a scientific knowing. A *rigorous* science of the *reality* of thought remains to be founded. And in order to constitute thought into a scientific continent, we have to find the means of displacing all the attempts at appropriation, exploitation, and division of its unique cause (of) the One.

2/ The current claimants share—this is the foundation of their competition—a certain division of intellectual labor and of its tasks. To Human Sciences, the science of thought—but a science without thought, whose object is simple "cognition." To philosophy, the science of thought— but a thought without science, whose object is a simple "representation" without the reality of the scientific object. On one hand, sciences that were created for other objects (whether human or nonhuman) have been projected on this new object; on the other, the object "thought" is hardly a real object, it is rather an aggregate of outlines, an assemblage of heterogeneous perspectives on thought. As always, when a new object tries to appear, one first claims to identify it through the incoherent combination of sciences created for other phenomena. This is why the *reality* of thought, its cause, has not yet been manifested as such or as the thing itself through a science *adequate* to it. It was simply outlined at the horizon of certain psychologies, of certain information technologies, of certain biological disciplines. How can we imagine that it can be reached through the accumulation of several different sciences that aim at objects other than it? On the other hand, philosophy claims (rightly, it seems at first sight) to be capable of situating itself at the heart of thought because it is identical to thought. But its problem is that it is not a science in the authentic sense of the term. It is not a rigorous description that does not modify its object in-the-last-instance; it is an autotransformative practice of thought. It is doubtful that philosophy can extract from what it calls representation the real kernel outside-of-representation. Thus some of the current sciences of thought are perhaps sciences (others are manipulative technologies), but their object is not thought. Rather: their objects are the objectivated and transcendent artifacts, lost in the World. And philosophy, which claims to access the essence of thought, does not do so via scientific procedures. At present, a science of thought is not possible, or it is divided in two. Thus human sciences and philosophy have in common the division of the nevertheless undivided thought. They have *a common problem, which stems from their transcendent presuppositions.* They have an external experience of thought, as object or as representation (ontological and constitutive of its reality). Obviously this thesis is meaningful only from the standpoint of another experience of thought, which takes place, if not more "internally" to thought, at least more immanently or more "radically." *This immanent experience of thought—where thought is neither an object nor a representation, neither a technique nor an operation of transcendence or of alienation for reaching this object—can describe itself in its essence and, on this basis, in its exercises and its works, i.e., philosophy and science.* It is science, insofar as it has its seat in the One.

PASSAGE TO THE SCIENTIFIC POSITION OF
THE PROBLEM OF THOUGHT'S ESSENCE

As soon as a problem of *simultaneous* distinction and unity (this is what we call "unitary") is posed, philosophy lays claim to it and takes it on: it is its problem, the task philosophy was created for. Science can thus claim to resolve this problem instead of philosophy—but in another mode and through another (nonsimultaneous) distribution of Identity and Duality—only if its essence has been elucidated and has shown itself to be anterior, from the standpoint of reality, to the essence of philosophy. This is the only way to displace philosophy's unitary claims, by means of something other than a "revolution." In elaborating a science of philosophy, our main objective is to recognize that science liberates itself from philosophical authority by its own means, i.e., by means of a rigorous autodescription in which it treats the One as an immanent guiding thread. On this basis it is then possible to recognize that science suspends the claims of various philosophies as well as their major claim to constitute the most originary experience of thought. This task of constituting thought, in the vastest sense (including "Reason" and "Intelligence"), into the object of science implies first that thought is wrested away from the illusions of the philosophical Decision. Only science can lift the philosophical resistance by manifesting it as such. So long as the attempts to unify its heterogeneous experiences remain *unitary* in style (simultaneity of Division and Unity) and take place by reduction to the authoritarian paradigm of philosophy, thought's continent will remain undetected as such. The discovery of the scientific paradigm's absolute autonomy, of the existence of an absolute original and primitive thought that is experienced as what traverses the process of production of scientific knowledges, wrests in one stroke all the experiences of thought from philosophy—even the philosophical, which becomes the object of a non-philosophical science . . . To rediscover and describe the real order that unifies—without reducing them—these qualitatively distinct experiences is then the second task, and it naturally follows from the first.

Such a project cannot therefore be realized by replacing philosophy with an *existing* "empirical science." This science would be more or less rigorously constituted; it would only be in this case a mixture of philosophical presuppositions, hidden psychological prejudices, and technomathematical procedures. We would thus remain in the current—cognitivist—state of the position of the problem. As we will see, the "sciences" and "technologies" of the artificial and of cognition are partially positivist subproducts of empiricist or rationalist philosophies, conclusions that are unaware of their premises. A science of thought takes up the cognitivist project again only as an indication or a symptom—as we have already said—and proposes to formulate it in its broadest extension, doing justice to the whole heterogeneity of thought (in particular to its philosophizing experience). An empirical science, like those that study cognition, would lead back to the current state of affairs, i.e., to philosophy's implicit and poorly elucidated domination over these problems. And this would create a supplementary vicious circle. The precise meaning of thought's constitution into a "scientific continent" resides primarily in the immanent seizure of scientific thought's Identity, in the way that this thought can describe it before its philosophical *division*; the grasping of its transcendental power to describe the real itself; finally, of its nonunitary function of ordering or unifying the fields of thought.

THE CONCEPT OF "TRANSCENDENTAL SCIENCE"

Thus a rigorous science of thought presupposes first of all a new, nonepistemological description of the power of truth proper to science. This description on its own constitutes a new science, at once a *transcendental science*, due to its real object (the essence of science) and its relation to the other sciences (it elucidates the transcendental structure in which all sciences, even "empirical" ones, already participate), and an *empirical science* alongside others, due to its need to discover its object-phenomenon outside itself. For if this science is constituted *in its essence* in the mode of an immanent autodescription of the *essence* of existing sciences, then it can only become an effective or "empirical" science when it finds its object or the contingent given that serves as material for its object (for example, in the philosophical Decision) and passes in this way from the state where it is a simple essence to the state where it becomes effective as "science of philosophy."

As a transcendental science, it is science (of) science (the (of) signifies "nonthetic (of) . . ."); it is the immanent autodescription of the essence of sciences. But as a science that finds an empirical object in philosophy, it adds a worldly and contingent given to its immanence and includes it under the conditions of this immanence. It is thus the same science that is science (of) sciences and science (of) philosophy, but according to a distribution of its identity and of its difference that is altogether distinct from their philosophical or unitary distribution.

From all points of view, it will thus be distinguished from philosophy. It will be a genuine "empirical" science; it will not be circularly code-termined *as science* by its contingent object; it will not result from the combination of a Decision and of a Position à la philosophical Decision. But like and more than philosophy, it will be a transcendental power of manifestation of the real essence of sciences and will not confine itself to constituting an empiricist and psychological description-theorization of thought. The current claimants to a science of thought—who are divided because they have in common the division of thought's Identity—are disqualified together in the name of what, from the outside, can appear as their synthesis in the Idea of a *transcendental science*, but is instead their Identity *before* its division. This Identity of a transcendental power of revelation in-the-last-instance of real essences and of their rigorous "empirical" description is an unacceptable paradox for philosophy, which seeks to divide Identity. But thought's economy can only be redistributed by a science—and we will have finally recognized that science is not merely a blind and technical process for the production of knowledge, but a power (more originary than the philosophical itself) of revelation of the immanent phenomenal givens, which form the "real essences," i.e., nonpositional Identities (of) self.

Perhaps we can specify philosophy's scientific impertinence with science's transcendental pertinence. The philosophical theory of thought is not the thought in which the ordinary man lives as an individual. It engenders—through its own operations of Reversal and of Displacement, of Decision and of Position—a certain image and a certain practice of thought. But this practice covers neither the thought of science nor that of "common sense" (as a type of experience). It imposes on them, by definition, its fully practical (decisional and positional) image and seeks to reduce all the other experiences of Intelligence to this image. It is the superior power [*puissance*] that is divided into itself and into science, into science and into common sense, which *proceeds* and returns to itself through them. It is the superior Unity of all these heterogeneous spheres of thought, whereas common sense and science are the deficient spheres of the philosophizing thought, which is not separated from itself in them without rejoining itself, rewilling itself, and so on, in order to be—*starting from their presumed ideality, i.e., from their presumed division*—a real and infinite unity, a unity for itself rather than for these spheres. In this case, the sense of the specific *reality* of science and common sense is altered: this reality is cut in half, idealized or unrealized into a degraded or deficient empirical reality outside philosophy, which must be sublated and critiqued, and into an identified thought, already identifying itself with philosophy, that affords it its essence, its sense, and its truth. Scientific thought is no longer real on its own and from its own depths, but through something else that has to make it intelligible in the mode of logos. It is understood by means of something else, which understands itself. Thus science is not grasped as such in its own essence; it is appropriated, captured, and reproduced as a "mystical result" (Marx). Philosophy has no end other than itself. This is why it transforms science's reality in order to try to adapt to it willingly or forcibly. It conflates the philosophical essence and the real as essence or the specific essence of the real. It is a voluntarist way of *unifying* and unitarily rubbing out the heterogeneity in the Idea, the Mind, the Will to Power, Substance, as well as in Difference and Différance, etc. Thought's real essence does not perhaps lie in its logical determination or its determination of logos, which is only one among others, which is particular even in its type of universality. Philosophy exhibits the real as autodetermination of unitary logos, but there is a "logic" proper to the real, to science, for example, and it is not the logic of the Concept, of Logos, of Difference, and so forth. Science has to be justified in its "empirical" existence, in a sense of this term that is not programmed by the philosophical Decision and can only be programmed through science itself. Science has its own experience and its own concept of "reality" and of the "empirical" (the "occasional").

Thus the solution that would turn science and "common sense" (the "ordinary") into simple empirical presuppositions—to be lifted, sublated, or deconstructed—of philosophical sufficiency is ruled out. Such "presuppositions" are always imperfect and empirical and have to undergo the labor and the sense-donating operations of the philosophical Decision. Here they are something other than "presuppositions"—i.e., necessary for the beginning, but ultimately inessential and can be "abolished" or "deconstructed," etc. Instead they form the *real basis* that conditions, at least *in its reality* if not in its effectivity, the philosophical experience of thought. It is not this experience that formulates the reality of these "presuppositions." The philosophical Decision is incapable of reproducing and truly transforming the movement of thought (for example, the scientific movement), which belongs only to itself.

THE UNITY OF MAN AND OF SCIENCE

This attempt intersects with a set of similar projects in the history of philosophy, but perhaps these projects failed—not as philosophies, but as scientific projects—precisely because of the excess of philosophy and the submission to the PSP. Marxism and positivism, for example, not to mention Kant, Husserl, and countless others. Consider—but this is an example, not an obedience—positivism. Between a science of thought and a philosophy of Intelligence like positivism, there are a few common points. We will mark the *appearance* of these points, since it is a matter of rediscovering the real content, the phenomenal sense of historical positivism. From our point of view, positivism is a philosophical success and a scientific failure because it took place on predominantly philosophical grounds and because it was unfaithful to science's essence in the name of a particular science.

• A "transcendental science" (not a transcendental philosophy as science) is positive insofar as it describes only phenomenal states of affairs. But it is transcendental because these phenomenal givens are strictly immanent. It is science's power to show itself as autonomous and first, as the power to elucidate (itself) as relation (to) real. This is the kernel of a science of thought, which is more originary than "empiricist" or "logical" positivism.

• A science of thought is, if you want, an (immediate) mode of the science of men, of the genuine human science we are looking for; it has nothing in common with a sociology and a psychology. It is the science of men within the limits in which, qua radical Identities par excellence, they are the cause (of) science, understood in the ante-epistemological sense. A science of men is necessarily a science of thought. This does not mean a science of reason alone, of transcendent and thus abstract intelligence. Rather, it extracts and describes the phenomenal real kernel through which reason and cognition access humanity and are immanent and real lived experiences of man as "cause." Under this definition, thought is man himself, but this is an anti-idealist definition. Thought does not codetermine human reality; it is man who, through his nonthetic reality (of) self, i.e., through the experience (of) self that he is, determines-in-the-last-instance the transcendent forms of thought that are grouped under the term *reason.*

• The task is to rediscover the real order of Intelligence. The recognition of the real order takes place via its "theoretical" rather than practical knowledge. On the basis of the real order (that of Identities and of Unilaterality), the continuous "progress," i.e., the production of knowledges, represents the only chance for a real and nonhallucinatory transformation of man's products (City, History, Language, Sexuality). We stay in the vicinity of Positivism (progress through order) on the strict condition of voiding it of any ultimate philosophical or transcendent reference. "Revolution" is a metaphysical and "terrorist" motif that promises order only in the form of terror and peace only in the form of war. But the real order in question is the one that the specifically human causality, tied to the *essence* of man rather than to his labor (activism and voluntarism), introduces not into the World, but from man to the World and to its authoritarian attributes. The real transformation of the World's transcendent or unitary orders can take place only on *the real basis of this nothing–but–human order that science expresses—in its essence rather than in its results* . . . Hence this general maxim: *to philosophy, order through disorder; to science, chaos through order.*

UNIFIED THEORY AND UNITARY THEORY: DETERMINATION-IN-THE-LAST-INSTANCE

This project is characterized by two complementary traits: 1/ it is a unified and not a unitary theory of the fields of thought; 2/ it is a scientific and not a philosophical theory of the unity of these fields.

Only a science can found a unified theory; only philosophy can found a unitary theory. We have to carefully distinguish between *unified* and *unitary*. It is not a matter of the same type of unity, although they are usually confused. A *unified* theory is not one whose unity is weaker than the unitary, more feeble and less consistent. It is more heterogeneous and more multiple; it is not less grounded or less rigorous.

Negatively, the *unified* is not a mode of the unitary, an empirical and weakened mode. It does not proceed by incremental induction and does not consist in gathering heterogeneous fields of thought from the outside. This is precisely the unitary style, the style of a general economy or distribution of knowing that proceeds through a *divided-dividing Unity*, through a *divided Relation* or *Hierarchy*, a *Coupling* of contraries (even a topological coupling). These modes of unification are excluded here: not only the massive metaphysical Unity of different types of "representation" that philosophy critiques, dissolves, dismisses, deconstructs, etc., but the more "differentiated," more "disseminated" types of unity it substitutes for them. The "unitary" is vaster than "Representation" or "logocentrism"; it merges with the philosophical operation of transcendence, decision and position, whatever its specific nature (Scission, Nothingness, Nihilation, Other, etc.). It is unity-through-division, ongoing-or-process-unity. The "unified" is a whole other thing.

Positively, the conditions of the problem are clear. Thought knows three or four heterogeneous forms. How can we allow them their specificity, in particular their specificity to the philosophical experience of thought, but also to scientific thought and, finally, to the technocomputing field? How can we allow them more specific autonomy than philosophy does (exclusion/interiorization, game, critique, differe(a)nce, etc.)? From this moment on, philosophy is only one thought among others (the same applies to its mode of unitary distribution). And since we cannot confine ourselves to a simple juxtaposition, the problem is to discover *a more liberal, more "phenomenological" and positive mode of unity that lets phenomena be without practically transforming them in view of a Unity (be it a divided unity). Is there a mode of Unity less unitary than philosophy and yet grounded and rigorous? And one that is compatible with scientific knowledges' chaos-of-the-last-instance?*

The Unitary is the correlation of a Dyad and a Unity, of a division or a distance, and of a Unity that is coextensive and superior to these two terms, at once internal and external to them such that each of the terms is accountable to it as a third party, even if it has not always been simply transcendent. One may recognize here the "Same" of contemporary philosophers. Another, simpler mode of Unity, the "dual," consists in abolishing this third and superior Unity as authoritarian, in making it useless, in leaving Duality to itself, in leaving only two terms—among them the One—to arrange themselves, without contracting any debt toward a third, without being accountable to an instance responsible for unifying them. Nevertheless, this Duality must not become a simple indifferent juxtaposition and an absence of thought. The solution consists first of all in radically "interiorizing," without remainder or transcendence, the function of Identity or of absolute reality that was divided between the third term and the Dyad of the first two: this third term finds its reason in itself, i.e., the One-in-One. Instead of imagining that it has to be identified with the Dyad in order to acquire some reality through the multiple, we will place this identity and this reality directly within it. This will be—we have already described it—an Identity-without-identification and a Reality-without-realization. In its turn, Duality necessarily changes its sense and loses the external and reflected reference to Unity. It becomes *a simple duality* that no longer derives from a scission; i.e., the second term then appears necessarily contingent for the first or the One. The true Duality, that which is strictly a Duality without third and superior Unity, is the one that forms this Identity-without-identification (and) the second term, but grasped this time in the radical contingency that the first inscribes within it. The phenomenal content, i.e., Duality's irreducible real kernel, is no longer the *two* or the *dyad* that presupposes a survey, a coextensive unity. It is the *dual*, the contingency of the second term thus unilateralized, a contingency that is itself transcendentally grounded in absolute Identity—in other words, it is not a juxtaposition.

We will call this matrix *a dual or dualitary rather than unitary unification.* Just as the "One" was an Identity without identification (it neither results from an identification nor produces one), the duality of terms is now a *Duality-without-division* or a "dual." On one hand, Identity is not obtained by reducing the second term to the first in a reciprocal and superior Unity, in a synthesis, a violent and reductive economy. It *lets-be* the second term (the philosophical Decision, for example, the field of thought that will be pervaded by radical contingency). On the other, the Duality neither results from a decision or a separation effected in a superior Unity nor reproduces such a Unity. It is not a praxis or a practice (of scission) exerted on a transcendent, already given Unity whose underside would be an identification. The second term is indeed—*in its contingency at least or in its relation to the first term*—grounded by the first. It is not indifferently juxtaposed to it; it is *allowed-to-be* as indifferent. This foundation of contingency does not reduce it (interiorization/exclusion, etc.) unitarily and hierarchically.

If science does not *intervene in* the contingent given, that is because in a certain way it receives this given from the outside. This formula must, nevertheless, be grasped clearly: it does not mean that science does not transform some givens in a technico-experimental way, but that this transformation does not pass through philosophical operations, through operations of the type "Decision" and "Position," "Reversal" and "Displacement," and that their interpretation by means of these operations (the theme of "objectivation") is an epistemological falsification. In a sense, and in this sense only, it lets be the transcendent given even better than philosophy does, which can only receive this given on the condition that it affects it with a "division" or a "decision" and, in a certain way, *constitutes* it in its sense or its being. On the other hand—this is the underside of the "letting-be"—science impresses a sterility of a transcendental origin on this given. The technico-experimental transformation of givens is not their *ontological constitution*. Even when they are inserted or included in the form of the "real object," they are not ontologically transformed. The consequence is that they are necessary only from the perspective of their function as "support" or "occasion"; for the rest, i.e., for the real essence of science, they are contingent and do not contribute to determining it in an essential or "ontological" way.

This operation of the One belongs to science. It is not only the operation of knowledge; it is a revelation of the real itself. But, far from being exerted in the mode of reciprocal determination (the manifestation of the real as its *pro*-duction as well, its *essential* transformation), it is exerted through a reflection, which leads back to the causality of the real as a determination that, at any rate, lets be the contingent term. This is the phenomenal residue—indivisible because of its minimal simplicity—of what was formulated a first time, in the materialist, philosophical, and transcendent context, under the name *Determination in the last instance.* Grasping the exact (and rigorous) meaning of this concept leads us back to the specifically scientific causality and to the "unified" or dualitary relationship. The description of Determination-in-the-last-instance brings to the fore the following traits:

• It is the causality that passes through the dual, through a "duality" that is no longer unitary or obtained by division, that is no longer continuous and reciprocal as are all the modes of philosophical causality, and that lets be the second term in its autonomy and its contingency and does not divide it in its turn.

• It is the causality of the real and not the causality of effectivity, the way in which the *real basis* acts on the transcendent and contingent given, the causality that presupposes in fact such a term.

• It is a unidirectional or unilateralizing causality; it prevents the given or transcendent term from *returning* to the real. "Last-instance" does not designate a first/last term in a continuous and circular series, but an irreversible order.

• It is a causality of the term as such, thus of a single term; it is not divided or distributed into two terms, which are equal from the standpoint of their hierarchy. It is the specific causality of *Identity* (the one that does not pass through a relation), the causality of the unique term or the *individual* in its solitude.

In general, we will avoid interpreting it in terms of transcendent, brute, or material causality. It is instead the transcendental causality that science exercises as such on the material to be known and whose correlate is a non-thetic-Reflection. It alone founds a science of thought, even more than a science of history. Determination-in-the-last-instance is the same thing as the *real, dualitary rather than unitary order*, the foundation of a new economy of the Continent "Thought." It makes possible a *unified theory* of the fields of thought. It has no relation, by its procedure, with a hierarchizing and conflictual unification, with the way in which, here and elsewhere, one will have attempted to coordinate neighboring "fields," to "throw bridges," to trace frontiers or adjoinments, to place all the modes of the philosophical Decision in a topology or in relations of proximity . . . What we have just described are the phenomenal givens of a *dualitary* thought, which nevertheless unifies its object in its own way, but with a minimum of violence. It does not, in any case, resort to the violence of philosophical praxis (Decision and Position, Reversal and Displacement).

When the time comes to describe a "science of philosophy" and its result ("non-philosophy"), all the concepts and results outlined here have to be taken up again and *specified* from the fractal perspective in terms of the object "philosophy."

SCIENCE OF SCIENCE AND SCIENCE OF PHILOSOPHY: THEIR UNITY

We can now exemplify the general structure of the "dualitary" unification in the case of the science of thought. We will certainly not "apply" an external and abstractly defined method. Science itself, as an autonomous paradigm, has supplied the model of every real and grounded relation. This is less an application than a description of science's effects.

It is obviously a matter of thinking science and philosophy together under the general rubric "science of thought," but they are perhaps not caught in a dyad in the way that philosophy considers and treats them spontaneously or as they are ordinarily imagined. In that case, their type of unity would be unitary once again; they would be again conceived in their "ideological" or philosophical image. The dualitary relation is applied to the relationships of science, of philosophy, and of cognition, only provided it has already transformed these last two experiences of thought into the object of science in terms of the Determination-in-the-last-instance. *The "dualitary," nonunitary duality is not the duality of "science" and of "philosophy" in general; it is the duality of science itself.* And it is the essence of science that grounds it. Thus we will not have a science and a philosophy face to face, but *two sciences in a dualitary correlation* through their respective object: a science of the essence of science (or of "empirical" sciences) and a science of philosophy. The first is a "transcendental" science, an episteme-without-logos, a transcendental epistemic rather than an epistemology. The second is the science (as well as a critique and a new practice) of the philosophical Decision in the way it is included in the first science, which discovers its particular object within it: philosophy itself. *There are two sciences here, but only because there are two types of "objects"* (science and philosophy are the two poles of every thought; cognition and the computing experience of thought are situated between these two poles), *for these two sciences are one only in-the-last-instance through their essence*, because it is the essence of science as such that makes them possible.

One should not see in this formulation the old unitary theme, as if (*unique*) science were *divided*, the way it is, for example, in Marxism between dialectical Materialism and historical Materialism: a dialectical or unitary formulation. Science, as we described it antephilosophically, is in fact an undivided Identity, which is only varied in exteriority and in a contingent way: through the contingency of this new object—philosophical thought and cognition. There is only one science, at least in the sense that the essence of science is anterior to every division and synthesis, whether dialectical or not, without being for that reason a transcendent (always divisible) *unity*. Just as the "dualitary" dual presupposes a second term, which is contingent with respect to the first, and is only a duality through this contingency that

makes it itself contingent (it would otherwise be unitary once again . . .)—
so science, far from being a divided unity, has a *dual* form. It is essentially
science (of) self and thus science (of) the philosophical Decision, *if* this
Decision at least is presented as a possible object. Such a "duality" does
not redouble, nor does it derive from, the duality that is immediately given
under the very epistemological rubric: "science *and* philosophy." It is not
accountable before a superior Unity, at once coextensive to the two terms
and divided. From the perspective of its conditions of reality or of its transcendental conditions, it is founded in the Identity-of-the-last-instance of
every science. And from the perspective of its objects and its extension, i.e.,
of the "effective" conditions of its duality, it is founded on the contingency
of the given philosophical Decision.

Thus, in terms of this new type of "relation" or rather of causality, not
only does philosophy no longer (ontologically) precede science, but even
the science (of) philosophy can no longer precede the science (of) science.
If there is a science (of) philosophy and if philosophy is thus subjected to
science (as the *object* of a science, nothing more . . .), it must proceed from
science. And a science (of) science must precede and include a science (of)
philosophy. We "pass" from the first to the second (without "leaving" the
former or "alienating" it) through the receipt of a new object; it is, no doubt,
very particular, because it has the power to give itself with the claim of being
other than science or in a state of resistance and self-defense against science.

These "two" sciences therefore correspond to the duality of the real and
of effectivity. But with one nuance. The objects of ("empirical") sciences are
phenomena that are extracted from effectivity, and sciences *include* them in
a "real object" specific to science. By the same token, the philosophical Decision is an effective given that must be in its turn included in the "real object"
or in the "form" of scientific objectivity, which is absolutely irreducible to the
simple transcendent donation of historico-cultural or physical . . . phenomena. But the philosophical Decision sees its importance increase. It will be
distinguished from other worldly or transcendent phenomena *both* (but it is
no doubt the same thing) as what claims to be the most general form of the
phenomena of effectivity or of the World and as what opposes the most vig-
orous resistance to science's extension toward it and above all to the "thesis"
of science's transcendental consistency.

So as better to describe the specificity of these two objects and their
"dualitary" relation, the freest possible relation, we can sketch out a

comparison with Marxism, with the relation it instates between science, as science of history (Historical Materialism = HM), and philosophy, as philosophy of the science of history (Dialectical Materialism = DM).

Like Marx, we will go from science to philosophy. Science is the criterion of every knowledge; we have to start from it in order to find philosophy. But this imperative, this guiding thread that we apparently take up, is still a vague and confused generality. In fact, if the science of history precedes philosophy in Marxism, it continues to do so in the classical reflexive and philosophical mode. Science (of history) remains a *historico-factual given*, the most certain science of reality (sociohistorical reality in this case), and philosophy (as DM) remains essentially a *reflection* on this science, on the possibility of thinking it, on its relation to the real of history, etc. From our point of view, this is once again the unitary construction and very general reflexive schema that make science an obscure nonthought, which produces knowledges but requires the *complement* of a philosophy—in this case, the complement of DM, which is responsible for "epistemologically" elucidating HM.

In relation to this Marxist version of the real order (from science to philosophy), the present terms should be apparently "inverted." In reality, we have to start from philosophy's equivalent, i.e., from the instance that fulfills the functions of the real cause and the transcendental. In our case, this instance is directly science rather than philosophy. And it is here a science (of) the essence of science, which is not yet created, but has to be. And we have to go toward a science of philosophy (rather than toward a crowning, terminal, and superior philosophy of science). But this time philosophy will be reduced to the state of a science's object rather than invoked as developing its own scientific virtualities or else as the founding philosophy of science.

So the trajectory is at once the same as Marxism's as well as "inverted" in relation to it. It is presented (at least from the perspective of philosophy as PSP) as the appearance of an inversion of the traditional relations of domination between science and philosophy. From our perspective, Marxism as always detected the right order, i.e., the *real order* between science and philosophy, an order required by the essence of science rather than decided or willed by philosophy. But it did not hold fast to this order. It hastened to reinterpret it with the help of the old philosophico-unitary conception, which subjects science to philosophy and strips it of every transcendental claim. Not only the real order requires that science be first and philosophy second, but two decisive clarifications must be added against philosophical resistance. The first is that science no longer needs a philosophical supplement in the form of a DM. It is nonthetic knowing (of) self because it is absolute knowing *or* knowing (of) the real and is thus not materialist (= philosophical *thesis*, for materialism is a transcendent form of realism, just as matter is a transcendent form of the real). Given the real-One, it is absolute or autonomous, the *equivalent*—just the equivalent—of a philosophy. This simply means: science assumes the transcendental functions of the pertinent relation to the real, but it assumes them in a non-philosophical mode, without needing to pass through the form of first Transcendence or of Decision. Since it is already, on its own, a thought in the definitive and full sense (but not in a reflexive form), it does not need a supplementary philosophy, a supplement of consciousness, reflection, meditation, or mediation in order to be what it is and be grounded. It finds its cause in the One and thus knows itself without passing through the operation of an auto-foundation. Science is a nondecisional process-*in-cause* (of) self.

The second clarification is that the science (of) science, which is first and precedes philosophy—and this is what allows it to replace philosophy in non-philosophical forms and a non-philosophical place—can no longer and must no longer be *only* a particular science, at least the way HM is *science of history*, the science of a sector or of a domain of total experience or of one *continent* among others. The science (of) science is not only the science of philosophy. Marx follows the classical reflexive and ontological trajectory: he starts out not from science *as such* or in its essence (this science is the object of the transcendent science or of first science), but from a particular or contingent science: if history and the relations of production are the *real* of socioeconomic foundations, and if HM is thus—this is clear—the science of a *real object*, it is still from our perspective a particular, empirical, and transcendent object. We have to start from science itself, i.e., from its essence, which, in order to be a "particular" object, is no longer one that belongs to effectivity and transcendence. It is the essence or the reality both of knowing and of thought in their undivided Identity. Marx takes the traditional path that goes from the particular to the universal, from *a* science to Philosophy [*la* philosophie] that will give the essence of this science. He does not leave the epistemological schema he tries to burst in vain. What we call the science (of) the essence of science is at once (1) a singular radical appearance or, as we call it, a chaos, which does not need philosophy in order to be real and not only in order to be a science of the real, and (2) an absolute "universal" experience in another, non-philosophical or scientific sense of the word. Philosophy's universality, not to mention that of the philosophical concept of universality, appears now as a "particular case" in the face of the theoretical power of Identity as cause (of) science. Philosophy becomes a simple object, a particular case for a way of thinking that, given its real individuality, no longer falls under the philosophical Decision, but is more powerful than it.

Thus it is at last the science (of) philosophy (the science of thought's transcendent forms and probably its "cognitive" and "computing" forms) that takes the place—another place . . . —of the science of history, of HM, with a supplement of generality, since HM is still dependent on philosophy. This is indeed what we sought: to form thought, its various fields, into a new scientific continent. But the science (of) philosophy and of Intelligence only replaces the science of history from the perspective of the studied "object," and certainly no longer from the perspective of the "subject" or the thought that takes place in that science. Philosophy is no longer the supreme legislator; the PSP is toppled. The philosophical Decision is a particular simple object that comes after history as the object of science.

Science enters into philosophies in two different forms, which are conflated as soon as science is interpreted in its turn in philosophical terms and as soon as its essence is forgotten.

On one hand, science enters into philosophies—it is true—under the regulation of the philosophical a priori itself and in view of this use. It is thus anticipated and programmed by the form it receives, which is the form of reciprocal causality. From this point of view, one knowing represents a potential philosophy for another knowing, and, reciprocally, a philosophy represents some virtual knowing for another. Two given phenomena, one scientific, the other philosophical, as distant as they may be from each other, form a priori a virtual connection that has to be actualized or fulfilled under the law of reciprocal determination or of nonseparability. This connection then gives place to a new philosophical sequence. On the other hand, science also intervenes in philosophy while remaining outside it or, rather, while remaining within itself as the real basis of philosophy. There is a double use of science: under the conditions of philosophical nonseparability and now under the conditions of what we will call a scientific *separability*. Duality of intervention effaced by the dominant philosophism.

Conceived in this second form, science represents the real that entertains a specific relation with philosophy, more exactly with nonthetic representation or the reflection in which this representation is from now on inscribed: a unilateral relation through which it can determine the representation without being, in its turn, determined by it. This unilateral or irreversible relation, that through which the One determines its reflection, is all that can and should be understood by "determination-in-the-last-instance." Even if this statement occasions misunderstandings, we will say that science takes the place of an infrastructure—from now on, real rather than material and transcendental rather than transcendent. The real, most universal infrastructure is the science of the essence of science and not the science of history in particular. But science cannot be reduced to its logico-theorico-experimental means, anymore than it can be reduced to some materiality. We avoid at once a materialism that is effectively "without thought," since it represents the denial of a philosophical position, and a scientistic positivism that would fold the *essence* of science back on its local procedures of representation and would thus conflate, by a new idealism, the real object with its representation.

Perhaps we can now better grasp what distinguishes science and philosophy, without any possible unity—at least a unity of the type of "synthesis" or "ontological difference." There is a fundamental posture of science vis-à-vis the real, entirely different from philosophy's posture, to which one habitually reduces science so as to better accuse it of having derailed and "objectivated" the good thought.

Philosophy produces effectivity rather than the real—a reality of synthesis and of simulacra that passes off in its eyes for the authentic real, the doubling of a first real which, it decides, is insufficient—to the same extent that science knows the real without producing it and produces only "scientific knowledge," which philosophy requisitions and transfers to effectivity. Philosophy grasps reality in terms of a coherent and unlimited system of operations of closure and opening, just as science grasps it in a "closed" or "finite" posture—a nonpositional posture in the most radical way—*and, on this basis, can finally give an unlimited technico-theoretical representation of reality.*

There is indeed a philosophical *domination* inside other possible dimensions or inside the possible dimension of other attributes (Language, Sex,

Power, etc.). But from this sphere of *dominance* in which the domination of attributes is exercised by turns, another sphere has to be distinguished: the sphere of determination or reality (which is not transcendentally conflated here with the forces and relations of production), in the grounded and rigorous sense of the terms, and in which science does not encounter philosophy. For, if there is a frontier, a mobile and ceaselessly crossed limit between philosophy and technology, *there is none between these two taken together and science.* Between them and science there is no frontier, but a duality, or rather a "dual" or an identity without identification or synthesis, without any possible unitary reappropriation; its unilateral relation is expressed in terms of "Determination-in-the-last-instance." The concept of the infrastructure, a transcendental concept, rigorously founded in the reality of its object, allows us to understand once and for all why it is in the end really irreducible to the philosophical "superstructure" and how every philosophical interpretation of science is an idealist reduction of the infra- to the superstructure.

The foregoing is a very schematic example of the work that a science of philosophy can carry out on its object and of the statements it can produce from this object, in this case from Marxism. Beside the generalization of Marxism, countless others are possible. Let's suppose, to conclude, that such a *scientific* generalization of Kantianism can be carried out. We will say that the instance of truth, the one that enables the delimitation of the philosophical illusion, is not science in its particular physico-mathematical figure. It is science in its essence, which should have been recognized as an absolutely real, transcendental essence from the start and by its own efforts. That science is, from and "in" the One, the authentic "transcendental subject"—as some neo-Kantians hypothesized, while conserving the partially transcendent philosophical and logical subject—is possible only if the transcendental nonthetic experience is finally recognized as real and not illusory, if it distinguished for that reason from every "intellectual intuition." But this is possible only if we finally manage to truly expel from it every transcendence and not to conserve, in the nonautopositional essence of the One, the least "parcel of the World" (Husserl). More than the transcendental Ego, which requires operations of reduction, the One fulfills this requirement as long as it is exerted as nonpositional testing (of) self.

By its essence, science is absolute and cannot presuppose or tolerate a supplementary "condition of possibility," a philosophical requisite. This discovery inverts the order of thoughts and foils philosophical objections. It is science that is the (transcendental or immanent) criterion of philosophy. A radical critique of the philosophical Decision becomes at last possible when the critical operation itself becomes secondary and is founded on the recognition of science's (transcendental) positivity, on the rock of a *real identity* that will never have been—as Fichte objected to metaphysics—at first formal or logical in order then to be "realized." Rather: an identity that is from the start recognized as real and "alogical."

The One is thus what absolutely disjoins science and philosophy, what dissolves their "epistemological" mixture, and what ensures science's precession on philosophy—their nonunitary unity.

If it is no longer science that "dreams" (Plato, Heidegger), then it is philosophy . . . These results can be read as a generalization of the Kantian critique.

6

THE CONCEPT OF
AN ARTIFICIAL PHILOSOPHY

FROM ARTIFICIAL INTELLIGENCE (AI)
TO THE IDEA OF AN ARTIFICIAL PHILOSOPHY (A PHI)

LET'S now approach the problem from the side of philosophy.

We define a trajectory here: the one that goes from AI and cognitivism to an A Phi. AI exists; it must be taken as such beyond the often premature, restricted, or insufficiently radical philosophical critiques that were leveled against it. On the other hand, A Phi does not exist, except as a problematic Idea. Under what conditions would an A Phi be possible? Taking AI, which has precise theoretical claims, as a point of departure, if not as a model, we seek to establish A Phi's theoretical conditions of possibility or, rather, conditions of reality. It is probable that this trajectory is not simple, that it is not a matter of effecting a transposition, a continuous transfer of the same technologies toward a different object, of continuously relaying the programming of psychological operations through the programming of philosophical operations. Two reasons are opposed to this. On one hand, the object to be simulated or reproduced (or even simply to be experimentally known) represents with respect to *cognition* a qualitative leap—a Decision—for which AI, in its very foundations, is perhaps not prepared. And, on the other hand, a reason in the opposite direction makes us think that AI is not—in its most intimate telos—what philosophy, what philosophical resistance, believes it detects within it in order to easily reduce it: a simple technological use of science. It may be that, under the name of AI, the scientific use of technology is also concealed, and this might change everything if science's essence, as we argue here, is wholly different from technology's.

Instead of proceeding to a simple extension and a transfer of technologies from Cognition to the philosophical Decision, a procedure that is only too obvious and too obviously "ideological," the only method is to isolate the various parts at play and in struggle within AI and Cognitivism. Before we recombine everything, a rigorous description—which is founded in a radical conception of the autonomy of scientific thought—requires us to proceed critically: by dissociating, at least provisionally, the heterogeneous types of knowledge that divide the field of AI. None of these *strands* (science, philosophy, technology) is reduced to the other, especially not the strand of science. This method is the only one that allows us to evaluate *what a machine can do from the perspective of thought.* What can a technology do about thought? This question will find a rigorously grounded response only if we begin by excluding some concepts that are less problems than fantasmic and premature solutions, like the concept of "technoscience," which *refuses* to analyze the respective roles of the scientific, the technological, and the philosophical. If, in opposition to this ideological way of proceeding, we "isolate" for example science in its real cause, within chaos in its solitude, then it is the economy of science's relation to technology and to philosophy that will be transformed by this "nonepistemological" thesis about science. AI's internal landscape will in turn undergo an earthquake that will remodel it quite differently and will make visible new potentialities of this discipline. There are, at any rate, several possible interpretations of AI in terms of the definition given of science and of its essence. So we can hope to pass from an AI to an A Phi—and in a way that is not a simple spontaneous generalization or a sterile and probably falsifying natural extension—only if we first take the trouble to reevaluate the respective relations of the philosophical, the scientific, and the technological.

Instead of immediately attempting a transfer from AI to A Phi and a simulation of philosophy, we will therefore elaborate A Phi's most general theoretical conditions, its inescapable conditions. We will explore its vicinities, its contours, and its possible models. This problem completely remodels the traditional relations between the machine and thought; it requires the calling into question of their specular fascination. This is particularly true if AI, as it seems, has to introduce a scientific break—scientific rather than epistemological—within a tradition, and if, on the other hand, thought ceases to be defined in a cognitive or cognitivist way in order to acquire its properly philosophical or transcendental form or dimension.

We must, first of all, clarify the "stake" of the (still problematic) project or *Idea of an A Phi*, which serves as our guiding thread. This stake differs completely from the uses that philosophers have for a long time been making of informatics. The current uses do not bear—as they do here—on the very Idea or Whole of philosophy taken globally (which is possible only on the basis of science; this isolation of the Decision and its invariants as objects of science grounds the possibility of an A Phi). Far from putting philosophy in its very essence in question and in play, they bear on *objects* it has in common with other forms of knowing (the text and its manipulation). Thus they bear on local or restrained objects of the field of philosophy, not on philosophy as such; and at the same time on overly general objects, on generalities. Philosophy's uses of informatics are both too restrained and not sufficiently specific. In fact, up to now, philosophers have used informatics for general tasks, which have not been specifically philosophical. Not even for a "textual hermeneutic" in the strict sense, i.e., the *essential reelaboration* of the philosophical concepts of hermeneutics and of the textual, but for textual givens that are deemed peripheral to thought. On this plane, the new tool makes it no doubt possible to partially eliminate the traditional theoretical dereliction in which the majority of philosophers—if not the most recent among them—have abandoned their own texts. But the intervention of informatics seems to be reduced to an *interpretation aid*, more than it serves to modify the very concept of interpretation. Aid to philosophical textuality: just an aid, and nothing more than an aid to the text.

More ambitious attempts to demonstrate or simulate the ontological argument, for example, or to resolve the defined problems were pursued. But these attempts to mechanically demonstrate philosophical thoughts are not related to the very essence of philosophy. In contrast, we postulate here the existence of this essence. In general, we can offer a few remarks on the concept of *A Phi* and its finalities:

a. Such a discipline, bearing on the essence or Whole of philosophy, on the philosophical Decision, can no longer be an intraphilosophical activity (founded on restrained presuppositions) that conveys and reproduces poorly elucidated ontological presuppositions. It can no longer tackle local problems exclusively. It must begin by calling into question what is understood by "philosophy," "science," "technology," and so forth. This critique presumes that we can isolate philosophy's *essence* in some way and dominate it through a form of knowing and technique that would not depend in its turn on philosophy.

b. The concept of A Phi requires then that we no longer have prefixed models drawn from AI (models of tasks, of procedures, of knowledges, etc.) and that we invent them gradually. In general, we maintain that if we are searching for an A Phi (and under this condition), then we can no longer posit, for example, the problem in terms of competition, performance, simulation, and reproduction of already fixed and finite tasks. This is AI's fundamental posture or, rather, the most general interpretation given of it. In this case, one would take a philosophical problem (the ontological argument, the dialectical Hegelian synthesis, etc.) and would provide a logical analysis or version of it, either algorithmic or heuristic, in order to program it. This impatience leads one to proceed tautologically by simply tracing AI's tasks and presuppositions. To reject this, we have to have a nonanalytical and nonpositivist idea of philosophy. We have to reject the existence of atomic or quasi-atomic problems that are susceptible to an algorithmic solution independent of their relations to others.

We should thus be wary of transposing AI's tasks and performances as such to philosophy. Otherwise, we would postulate the homogeneity of philosophical problems to those of cognitive psychology and those of logic. It may be that the entry of philosophy's essence on the scene will *subvert the very nature of the tasks that one would assign to an A Phi*. We should not be obsessed from the start with questions of the type: what purpose should an A Phi *serve*?

- aid to the reading of texts (which reading?);
- aid to the interpretation of texts (which interpretation? which textuality?);
- aid to the philosophical decision (what can this mean?) and in the diagnosis of "philosophy" that deals with such a statement;
- simulation and production of philosophical quasi systems (how does one recognize a system as "philosophical"?);
- finally, "rigorous" demonstration (or invalidation) of some famous metaphysical argumentations . . . , etc.

This *position of the problem* (what purpose does an A Phi serve, which *aid*—demonstration of arguments, creation of systems, etc.—to the philosophical Decision?) must be suspended. In fact, the only perspective that will authorize this suspension and, at the same time, respect the autonomy of the philosophical Decision, without imposing an *empirical* reduction on it, will be the perspective of a *transcendental science* whose principles and conditions of reality were posited earlier, a science acquired by non-philosophical paths and thus capable of being a science of philosophy. The Idea of an A Phi is a milestone on the path that leads to this science.

To all these questions, only philosophy can in some sense respond, especially if they were elaborated and posed on philosophical presuppositions, as is inevitable, in a sense. In an A Phi, philosophy cannot be a simple passive object, since it intervenes, in any case, in the position of these tasks and objects. We will not say that these tasks must not be proposed to an A Phi. We will only say that the transfer of procedures, of levels of tasks and the very idea of task, the specular context of competition, competency, and performance, postulate the reduction of philosophical problems to cognitivist models of tasks, while the former already act in and are broader than the latter. An A Phi that reaches the height of philosophy, of its complexity and of its essence, can only transform the very tasks that one proposes to it and transform them in terms of the invariants of the philosophical Decision. It will generally refuse to proceed by analogy.

What legitimates *at first sight* the hope of an A Phi is AI's increasing orientation toward the problems of life, everyday experience and common sense, toward problems of obscure, blind, and even subconscious reasoning. These problems of common sense, good sense, ordinary practice, and everyday usage are very close in their complexity to the problems of philosophy, much closer, at any rate, than logic, at least so long as we do not have an a priori "scientific" image of philosophy, which almost always means, in this case, logical, atomistic, and behaviorist and "rationalist" rather than real or transcendental. In this direction we can imagine an *expert system* whose goal would be to simulate and use, to assist, above all, *the philosophical knowing*, which, even when it is not the knowledge of an expert in philosophy, always has (in itself, at least) the nature of an expert knowledge (concrete, overdetermined, intuitive, quasi unconscious). This is why it is particularly difficult to formalize and represent it in its philosophical concreteness.

Thus we only have a guiding thread at our disposal: the *Idea of A Phi*, which leaves the situation open. We cannot constitute an A Phi by addition, accumulation, transfer, or analogy. We will have to invent it and not to trace it from what already exists. This is why we have to engage in the *process* of elucidating its theoretical bases, a process with multiple phases. And perhaps the final result—what we called an A Phi in a slightly "decalcomaniac" way—will hardly resemble what we imagined at the start.

ELABORATION OF A PHI'S CONTENT AND TASKS

This formula covers a triple program: 1/ the inventory of the traditional critiques of philosophy against AI; 2/ the description of the spontaneous philosophies that sustain AI; 3/ the—problematic—extension of AI toward philosophy, the Idea of an "artificial philosophy" (A Phi). What grounds this program, which is inscribed in the vaster program of a *science of philosophy*?

Instead of describing AI's codified practices, we will seek its intimate goal, its telos, in view of extending up to philosophy what is only outlined within it. This telos seems to be the following: AI corresponds to a scientific "rupture" or "revolution" in the problem of a science of thought, an experimental science with a technological basis. A whole other thing, by consequence, than recipes for simulating thought. This rupture has precise historical and mathematical conditions, in particular the invention of new logical, mathematical, and technological means that allow thought to be *reduced* to reasoning and reasoning to arithmetic. This rupture defines an upstream and a downstream. Upstream: the old philosophical and fantasmic project of a (specular) simulation of thought by the machine as well as a simulation of the machine by thought.

The equation thought = machine has to be understood as a restrained mode of the great founding equation of metaphysics and of the philosophical Decision in general: thought = the real; logos = being. It would be easy to demonstrate that the equation thought = machine underwent, on the plane of its most general presuppositions, exactly the same vicissitudes and the same history as its metaphysical matrix; that it is capable of receiving the same interpretations; finally, that it implies the same specular and circular fascination between its terms, which are more or less deferred or delayed depending on whether one moves toward contemporary interpretations and "deconstructions" of this equation. The dyad thought/machine can be deconstructed exactly in the same way as any other dyad. Generally, the reduction of the problem of the thought/machine relations to the matrix and invariants of the philosophical Decision is a preliminary task, which is absolutely necessary for starting to pose this problem correctly and to discern the inevitable Greco-occidental aporias within it.

Yet AI is announced as the chance of finally breaking this circularity, in which no term manages to determine itself and oscillates in an aporetic and amphibological way between its identification with the other and its repulsion. This is an attempt—more scientific or more technological in spirit, this must be examined carefully—to place the problem of these relations on a controllable and experimental terrain and to put an end to the game whereby thought and machine are mirrored in each other. With AI, this is at least the project of putting a science in philosophies' place and thus changing the terrain, establishing a new type of rigor and a new type of relation to the real. Philosophers have to take this attempt seriously, if only because it proposes to remove from philosophical legislation one of its most fundamental objects or, better yet, the most general condition of a relation to the objects: intelligence or reason. In effect, under the name of "general intelligence," of "intelligent action," Cognitive Sciences (even beyond AI) propose nothing short of constituting a *science of reason* and perhaps even a science of thought; an experimental and technically armed science. This is their long-term ambition, their telos, and, from this perspective, they intend to assume the whole heritage of Human Sciences and to present themselves as their crowning. In the face of what is necessarily a danger for them, philosophers should stop believing that AI contains more artifice than intelligence. To believe that every intelligence can only be artful and Greek before being artificial and informatic is not perhaps the best means of fending off this danger.

This is why AI's downstream—the scientific downstream—is more important than its philosophical upstream, provided we consider this scientific downstream in the new relations it can establish between itself and philosophy. If we extend AI's and cognitivism's telos a little further, and if we return to what they merely outline, it seems that their project can be radicalized, transformed but enlarged. We could, for example, consider them as sections of a cone whose base would from now on be philosophy itself, the philosophical Decision and its invariants, and no longer the simple *cognition* that is probably (at least in one of its aspects) nothing more than a restrained and transcendent concept of philosophizing reason, and whose opening angle or peak would no doubt be science, but, on one hand, nothing but science instead of the current blends of technology and science and, on the other, the Idea of science instead of the blends of particular sciences like logic, neuroscience, or cybernetics, particular sciences that in any case—and this is why we raise this objection against them—are treated as the paradigm of every thought and function surreptitiously with the transcendent claims of an *essence* of philosophical reason. Under the name of *A Phi*, which serves as our guiding thread, we thus try to clear out a possible path that would go from AI as it exists to an authentic *science of the most deployed thought, i.e., of philosophy*. Philosophy, at any rate, would only be one dimension in a science of thought whose pivot would be a science of transcendental, albeit non-philosophical science. A *science of philosophy*: such a project rules out that it is a new (Husserlian, Cartesian, etc.) avatar of the *philosophy of philosophy*, i.e., of the project that is partially realized in the history of philosophy.

We will be wary of unilaterally critiquing AI as philosophers (especially continental ones) often do. On the contrary, we will take it as a symptom to be analyzed and displaced, rather than as a ready-made model to be "transferred" or dogmatically and unduly extended to the philosophical Decision.

How should we proceed? In two ways. We will treat AI and Cognitivism first as compromise-formations between philosophy itself (its essence and its most general invariants) and some poorly elucidated philosophical presuppositions that serve as their basis. So much so that we will seek not only to critique, but to *affirm* them and to show that they deal with the essence of philosophy more intimately than they themselves imagine.

We will therefore treat them as another compromise-formation—in a new sense that remains to be determined—between their essence-of-science (the science (of) science presumes that science has an essence) and their own understanding of science, their auto-interpretation as scientific disciplines, their will or desire to present themselves as sciences.

Put differently, the strategy we pursue to "pass" from AI and Cognitivism to A Phi consists in opposing twice (to their autocomprehension and to their desire to be philosophy and to be science) what is considered here as the radically and rigorously thought *essence*. This essence is presented in two heterogeneous forms: the essence of philosophy and the essence of science.

1. To their sedimented philosophical presuppositions, unrecognized as such, and thus to their restrained autounderstanding of philosophy, we will "oppose" the essence of philosophy in order to transform these disciplines from this perspective and to render them coextensive with this essence, to deliver them from their ever illusory and restrictive (empiricist and rationalist) philosophical autointerpretations, which misjudge or repress the deployed essence of the philosophical Decision. The procedure consists in making visible, from inside and outside AI and Cognitivism, philosophy's full and specific requirements, in particular its transcendental dimension of "decision."

2. To their autointerpretation as sciences, in which they are thought as mixtures of empirico-rationalist philosophies and of particular empirical sciences (logical and mathematical science, neuroscience, theory of information, etc.), we will "oppose" a more radical concept and, in any case, the "thesis" of the transcendental consistency and autonomy of science-as-thought. Given its immanence (the criteria of scientific *thought* are nothing-but-transcendental criteria rather than empirical or rational), this concept, which is not acquired by philosophical means, but originarily from itself, will constitute the most difficult ordeal [*épreuve*] for them and will allow us to evacuate the surreptitious mixtures that are formed from philosophical empiricisms and/or rationalisms and from the essence of science.

We have in fact shown that there is an essence specific to science; a quasi "ontology" proper to sciences—at any rate, a transcendental thought that does not fall under the authority of philosophy and does not engender an epistemology.

What we will create will therefore be neither a practice nor an epistemology of AI. We will ask: under what conditions can AI *become* a rigorous science of intelligence deployed in its ultimate possibilities, i.e., a science of philosophy? We thus propose is to *inventory the conditions of theoretical production of a science of philosophy—or of a ("full" or "enlarged") science of reason, starting from restricted models of AI and of Cognitivism.*

The fundamental condition is to restore science's autonomy vis-à-vis every epistemological recuperation and thus to proceed to something other than an epistemological "rupture" or "revolution" in relation to science. AI suffers in its development from overly limited and encysted theoretical bases, as much on the scientific as on the philosophical plane. The passage to an A Phi entails disrupting AI's internal economy (sciences, philosophies, technologies) and, above all else—this is the condition of everything—the economy of the relations between science and philosophy.

The implementation of this strategy assumes that AI and Cognitivism are treated in a complex and differentiated way.

• AI is apparently, and initially, treated as a model (perhaps a not yet elaborated but extant model) of a technical and scientific discipline of intelligence. In effect, it is necessary to yield *first* to appearances, i.e., to take AI and make it take another step, to extend AI to philosophy itself, to proceed experimentally and to see in any case—even if this is not enough—the irreducibility that subsists in the philosophical Decision that is subject to the informatic test. This extension of AI's procedures toward the ultimate mechanisms and operations of the philosophical Decision (beyond a simple simulation of local reasonings and abstract statements) risks constituting a radical trial for philosophy (something like *the heaviest weight*). But it is inevitable for two reasons. First, it is better for philosophers themselves to intervene in the management of this experiment, which concerns them directly, but provided they know how to ascend to the last requisites of the philosophical Decision and do not confine themselves to shouldering once more, and naively, this or that "system," which is adopted without prior examination. Second, it is inevitable that AI does not check its ambitions to simulate natural language, the most general mechanisms of reasoning, and, lastly, the specific knowledges of experts. A last ambition necessarily looms beyond "expert systems"; we must term it Artificial Philosophy (A Phi). Moreover, this ultimate extension of AI is not necessarily a matricide, even if it has the aftertaste of a matricide. It would for example be a way of experimentally verifying Heidegger's thesis about the affinity of "metaphysics" and "technology"; A Phi represents in some sense *the absorption in itself of occidental philosophy, which recovers itself through its ultimate technical possibilities.* The fundamental problem of this first operation is to fix at its right level—i.e., at the highest level, that of the philosophical

Decision—the trial to which AI is for its part subjected, in other words: the exact nature of philosophical thought's operation, which it should be able to simulate or reproduce. From this point of view, there will be no dearth of naive individuals and hasty enthusiasts who, failing to push the analysis of what a philosophical Decision *is* sufficiently far and confining themselves to a superficial description of philosophy, will believe they have finally invented the philosophical robot . . . It is probable that the principle of these illusions is the following: one stops the analysis of the philosophical Decision too early, fixing it on the system of logical rules it can contain and constituting them into "negative" conditions of philosophy's exclusion. On this basis, it becomes easy (even for a child) to establish a program of rules that prescribe the rejection or the avoidance of this logical code. But if the philosophical Decision is the rejection or avoidance of something, it is not primarily the rejection of logical rules; it is the rejection of ordinary and everyday language. And it is much more difficult to formalize the negation of everyday language (for example, of common sense) than the negation of logic's rules, especially when this negation is no longer merely logical but also *real* or *transcendental*, as is the case in every philosophy that rigorously thinks its essence. It is tedious to have to remind each new premature attempt of this kind that the philosophical Decision does not obey (or reject) some rules without destroying or rejecting others (or without inventing others). It is this transcendental mechanism that AI should be able or unable to reproduce. All the rest is a phantasm.

• But this first operation is still an appearance, even if it is objective. If we want to pose the problem of a massive intervention of informatics, massive and principial [*principielle*] in the most fundamental mechanisms of the philosophical Decision, in the very operation of philosophizing or thinking, then AI has to be taken, not as a model to be incrementally extended and generalized in an abusive or "ideological" way, but as a *symptom* to be analyzed and displaced. No doubt as an index of tasks and a reservoir of procedures and inventions, but perhaps also as a symptom, that of the relations between informatics (i.e., technology combined with science) and philosophy (which is simply presupposed in this case). Truth be told, if we exhibit AI's philosophical presuppositions (this exhibition is, at any rate, necessary in view of an A Phi), we should also exhibit its scientific presuppositions and flatten all these components. The path toward A Phil will be arduous and complicated.

A PHI'S STAGES

Under the name *A Phi*, we can imagine several practices that we have to select from—three concepts: A Phi I, II, and III.

1/ *A Phi I*: a simple extension of AI (of its procedures, its finalities, its concept of tasks, its *autointerpretation* in general) to philosophy. A Phi is then an extension and an outbuilding of AI. This is a *pre-dominantly* technological, rather than scientific interpretation of A Phi and, to be sure, of AI before it. Techno-logo-centrism is not exclusive within this interpretation; it is dominant. This extension of AI is spontaneous and stems from its auto-interpretation as *technoscientific mixture*. It is a rule that AI and Cognitivism, left to themselves, would interpret themselves starting from science and technology and would consider them united in a new discipline. From our perspective, this autointerpretation of AI and its spontaneous extension to philosophy are specular theoretical effects of the technoscientific mixture in the mode in which it is ordinarily practiced, i.e., of theoretical interpretations that stem from its dominant technological side. In other words, the technoscientific mixture is reproduced in the theoretical mode as mixed or determined essentially by its technological side and understood in a nearly exclusive way through it. This mixture is not then rigorously analyzed—we should say "dualyzed"—on the basis of the inequality of science as determinant and technology as dominant. AI's immediate and spontaneous extension to philosophy stays within the symptom and is confined to reproducing it without taking the trouble to analyze it. We may recognize here the majority of attempts at A Phi that have already been undertaken or will be undertaken. The result is a "restrained" A Phi, in which the devalorizing models of existing AI are extended to the Philosophical Decision (PhD) and falsify it; and, as an effect of this technologization of the PhD, we will have not a scientific, but a technological limitation of the possibilities of the new discipline.

2/ *A Phi II*: an extension of AI, no longer as a technoscientific mixture that interprets itself *more* as technological, but as *uniquely* technological. In a sense the mixture begins to be shattered or dualyzed, but for the benefit of the most affirmed techno-logo-centrism. It is philosophy, in its Nietzschean or Deleuzean form for example, that attempts or *could* attempt this philosophical autointerpretation of AI's technoscientific mixture as purely technological (the reduction of science to technological connectivity) and attempt to extend this interpretation to the PhD. Obviously the result would be tautological. If the pure technological syntax is conflated, as it seems, with the PhD's syntax in the Nietzschean regime (finished or autoaffirmed metaphysics = finished techno-logy), then the simulation by AI, understood as absolute technology, produces no gain. And, conversely, every the PhD is already an artificial autosimulation: PhD = A Phi; every philosophy is artificial. This result can also be read as a generalized A Phi in the technologocentric mode. In general, philosophy is no longer an empirical machine, but is exalted as an overmachine.

3/ *A Phi III*: an extension of AI whose mixture is now "dualyzed" on the basis of the primacy of its scientific side, which is no longer understood technologically and/or philosophically but starting from itself, or as the *real basis* of the technophilosophical superstructure that belongs, at any rate, to AI. This is the longest path. It no longer corresponds to a spontaneous, uncritiqued, and "ideological" extension, which leads to a technological overinvestment of AI. Instead, it leads to a transformative extension of AI itself before constituting an extension of the PhD. This transformation is complex and has been sketched out. It produces the most innovative and radical A Phi. Nevertheless the result will be less an "artificial" philosophy—it is a difficult not to think of this expression as a simple metaphorical extension, i.e., still under the technophilosophical authority—of AI than a *science of philosophy*, which will use the technological representation of philosophy as a scientific and no longer technologocentric ingredient. What will thus be founded is less a philosophical avatar of AI than a *pre-dominantly technological science of the PhD*.

This third solution thus makes science intervene twice and no longer a single time, as is the case in AI's autointerpretation with the help of the concept of "technoscience":

• A first time in the use of "reduction" exercised on the PhD, a use that is still philosophical in nature. Science is not exhausted in this function, but we can always consider and draw from it a particular use, the one it acquires when it is technologically realizable, programmable, or usable within the limits and under the domination and anticipation of the *technological a priori*. This technological side of science (not its technical apparatus, but its use or its investment in technologies) can be easily interpreted in philosophical terms, in terms of the PhD (for example, the Cartesian, Hegelian, Nietzschean . . . techno-logics), as the technological itself. Moreover, it can be used for the purposes of *reduction* ("transcendental" or not) in the same capacity as any PhD vis-à-vis empirical phenomena. Applied from this angle to the PhD itself and as such, this use of science is the equivalent of the PhD's autoheterocritique. It quickly reaches its limits and slips into tautology, since the power of science is then the very power of philosophy. The philosophy of philosophy and the technology of technology extract only their own invariants; they *auto*extract themselves as invariants. This use of science in terms of a (techno) philosophical reduction is not very fertile, but it is the only use authorized in A Phi I and II, and especially in the latter.

   • A second time in a scientific-transcendental use, which is always exercised on the PhD, but such that the type of science's transcendental truth is no longer, as we have seen, the one that philosophy knows. Already, this "nothing-but-scientific" use no longer represents on its own a first reduction or a first technophilocentric use of science. Furthermore, science in this use, which is nothing-but-transcendental and is also not "meta-physical" or techno-logical as in the preceding case, can act on the PhD without insidiously exploiting its procedures and operations, without simultaneously borrowing its essence. Its nothing-but-scientific efficacy can be formulated as an actual or already complete reduction at the moment when the PhD is affected by it and attempts to resist it by thinking it in its own way.

   In its practice and its current theoretical bases, AI is a technologocentric autointerpretation of science, a technophilosophical requisition of its essence, a science that is required under technoexperimental conditions and manipulative finalities, an attempt to manage thought, which is founded on a prior management of science. AI responds to the criteria of science, of scientific *thought*, only through the criteria of technological production—it is a philosophy disguised as a science by way of engineering.

   With A Phi III, we are dealing with a whole other thing: instead of applying to the PhD a science that is already interpreted by technophilosophy and that already presents itself in more or less restrained and dogmatic or enlarged and "thinking" autointerpretations, we will apply to it a science reduced to its essence of science and its transcendental power to manifest the real through the production of knowledges.

Whereas A Phi I and II are subproducts of AI, i.e., of science's technological requisition, A Phi III corresponds to a real surpassing of AI toward a *science of the PhD*, a science of philosophy as much as of technology, and in any case a founding discipline of every subsequent attempt to simulate or artificialize philosophy and no doubt "cognition" as well. From this point of view, the science of philosophy will be to A Phi what science is (really, outside every technologocentric autointerpretation) to AI.

A science of the PhD will thus pass through the two typical phases of A Phi III. In relation to an explicitly transcendental philosophy like Husserl's, what we call science or the nothing-but-transcendental no longer needs to proceed through an *operation* of reduction. Only the philosophical use of science turns science into such an operation. So much so that if we distinguish, as in Husserl, two levels of science's intervention, they will not have the same content and will be shifted a notch in their relation to the levels of "Phenomenology."

Whereas in Husserl science, requisitioned as philosophy, culminates in a transcendental reduction (limited to certain philosophical positions and not yet extended to every philosophical positionality and decisionality), here science, which is no longer philosophically requisitioned, *begins* from the start through the efficacy of an already actual reduction. And, even as "nothing-but-science," it no longer proceeds through an *operation* (decision-position) of reduction. A *science of philosophy* (and therefore an A Phi founded on this science, i.e., an A Phi that uses at the interior of scientific representation the technological representation of philosophy) is no longer, properly speaking, a classical *reduction* of the PhD; it is its "dualysis," its in-differentiation, its *unilateralization*. A reduction would remain circularly caught in what it must reduce—this is the common lot of the PhD—but science has at its disposal an other-than-philosophical causality on philosophy. Philosophy, as the simple material of a science, is included in the "real-object" or subjected to the "objectivity form" peculiar to science; it ceases to be considered only in an empirical way, from itself in this case. To this material (the PhD or a particular PhD), a technological simulation can already belong, an "artificial" representation or production, local and perhaps even global, but it will no longer be determinant of the very essence of scientific objectivity as is the case in A Phi I and II.

Seen from A Phi III, these three solutions represent stages in the path that goes from AI to a radical science of philosophy and of cognition. The point of departure is the technoscientific mixture of current AI, an auto-interpreted mixture in a technophilosophical and thus not-yet-dualyzed mode. A Phi I's attempts and "results" can eventually be included as materials in this point of departure.

The second stage begins with the analysis of the mixture or with the dismembering of AI into its components. But this analysis is carried out from the perspective of technology alone and presupposes its authority over the whole whose unity—a technological or unitary unity—is reconstituted at science's (now complete) expense. Here again, A Phi II can be integrated into the materials and is no longer but one stage in the production of A Phi III's concept.

Finally, the last stage dualyzes AI's mixture by constituting science, no more as one "side" of the mixture, but as the *real basis* of the technophilosophical "side," the basis that founds a science of this side. A Phi III (what we call non-philosophy, in other theoretical and more purely philosophical contexts, or the New Technological Spirit in theoretical and more purely technological contexts) is not the *truth* of A Phi I and II, the stage that would recapitulate them in a superior unity. It is, on the contrary, their radical "analysis," their *dualysis*, their manifestation as symptoms, and this manifestation is enough to denounce them as symptoms or mechanisms of the self-defense of philosophy and technology against science. This dualysis of AI amounts to reinscribing it into the essence of science, into their special "relation," which is the relation of determination-in-the-last-instance of the AI by science.

AI appears then as a restrained, autointerpretative—and thus predominantly technophilosophical—form of A Phi III. In order to access A Phi III, we have to pass through the science of philosophy/technology, i.e., we have to first resolve the crucial problem of the unity (of the type of unity) of technological causality and the specifically scientific causality (of science *as such*, not of this or that local knowledge) in a "real" or concrete machine. We can only resolve this problem *by giving a determination-in-the-last-instance of technology by science, rather than a technologist interpretation of science.*

The general strategy of this attempt resides in a double restitution of the essence: return from the restrained concepts of the technological and the philosophical to their essence and return from the technophilosophical interpretations of science to the essence of science. *A Phi III is then the new "synthesis" of technophilosophical causality and of scientific causality.* Is it a matter of a special "machine" (in the intratechnological sense) capable of this combination? Or, rather, is it a matter of a structure that combines the real basis of science and the most powerful machine that exists, the PhD as autoaffirmation of the machine or as "overmachine," which uses (beyond informatics) information that is neither continuous nor digital, but the synthesis of the two? Such a machine could "simulate" the technologico-philosophical as such, and therefore every local, less universal, machine or the PhD—this would in fact amount to an autosimulation. But it can only do so on condition that it and its operation are included in a *scientific representation* of the technological or of the philosophical and that their autosimulation is subjected to the enterprise of their rigorous scientific description.

This double extension must be understood as follows. On one hand, the rational is represented by the technophilosophical side. A machine is rational; it does not necessarily have a universal rationality, but it has a local and restrained one. The only truly universal rationality is the philosophical and/or the technological. There are thus multiple and distinct a priori forms of rationality, which do not necessarily coincide: rationality of a machine with logical programming, rationality of the brain, rationality of games, etc. The simulation is not necessarily possible from one rationality to the other, from one region to the other or from one mode to the other of the PhD *unless it passes again through the PhD itself as the invariant essence,* which is the most univocal rational.

On the other hand, we have the side of the "real" or the *real basis* of A Phi III: it is science, which relies on reason, but is not reduced to it by its essence, which uses the technophilosophical by determining it in-the-last-instance. A Phi III is therefore not an arbitrary machine. It is what a machine that simulates thought becomes when its power of simulation is radically potentialized until it turns into a general autosimulation and when its scientific basis is phenomenally exhibited as such. Instead of a machine or even an overmachine that simulates thought, *A Phi III is a science of thought that uses the technophilosophical autosimulation of thought as a simple procedure of scientific knowledge.* It is less a machine or a technology of thought than a pre-dominantly technological science of thought. It can only be controlled technophilosophically—rather than logically—by its "superstructure" of technophilosophical representation. By its real basis, it does not pertain to an operation of control because it is transcendental or *index sui*. Once more, A Phi III can only be reached on the basis of the rigorous dissociation, the dualyzation of the technophilosophical and the scientific as two heterogeneous essences: that of ontological objectivity and that of transcendental reality. The concept of "rational machine" expresses their confusion as much as their positivist flattening, in the same way as the concept of "technoscience."

These clarifications constitute the grounds and limits of the concept of A Phi in its most rigorous and most complete form. We have to maximize AI's grip on the PhD and on the whole surface of philosophy's objects and operations; we have to potentialize its technological and philosophical procedures while knowing that, from this first point of view, A Phi encounters a de jure yet mobile limit that it can only displace without destroying. This limit is internal: the PhD itself as relative-absolute limit or as invariant (tendency-limit). But AI's extension and potentialization into A Phi discovers—outside it this time, but inevitably—an absolute limit, i.e., something other than a limit: a determination-in-the-last-instance, a causality of science on the technophilosophical side. Once the philosophical critique of AI is deployed, once it is extended to the essence of the technophilosophical, it is then necessary to extract AI's real-transcendental or scientific kernel and to renew it in its essence as well. This double operation, which is asymmetrical by its procedures and its finalities, founds *a rigorous science of philosophy and of cognition.*

We still have to show that Generalized Fractality allows us to resolve the problem of a science of philosophy and thereby of an *Artificial Philosophy.*

7

THE FRACTAL MODELING
OF PHILOSOPHY

THE PHILOSOPHICAL CLUTTER AND ITS SELF-ERASURE

AS a certain regulated—regulated as much as deregulated—use of natural language, philosophy quickly appears as a clutter, a manifold of motifs and objectives, objects and themes, syntaxes and structures (more or less mismatched or poorly assembled), a deformable tissue of aporias, of unmasterable neighborhoods, a particularly irregular curve. Postmodern thought is the exploitation of this irregularity, which, as we have already said, is only semifractal and weak; it is the exploitation of this fragmentation, which is not without rules, but is without rules that would not be themselves mismatched and hesitant.

Nonetheless, it is only a matter of a semifractality and of a curve. Let's recall some results of philosophy's dimensional description. This description brackets philosophy's objects or contents in order to confine itself to its syntax of decision and position; on the other hand, it does not trace the dimensionality of a system of the number of philosophy's terms or of its empirical "dimensions." It can be said of any philosophical decision that it possesses either (a) essentially two "principles" (One and Dyad); (b) 2/3 terms (the two opposites of the Dyad and the One that is partially identical to one of them); or (c) precisely a dimension—in the scientific sense of the term—equal to 2/1 or 3/2, depending on the analysis, but an intrinsically fractionary or (semi)fractal dimension that tracks the empirico-transcendental distinction; this distinction is not found in this form within scientific knowledge. Now if we vary the degree of resolution of our gaze on the philosophical decision, either plunging by "magnification" into the details of structures and objects or, on the contrary, generalizing and extracting invariants by distancing and overhanging, then very different properties of philosophical thought emerge, seemingly heterogeneous organizations and logics until— depending on the case—the complete erasure of the structures in the details or the erasure of the details in the invariant structures. And yet it is remarkable that, from one end to the other of their path, these scale-changes or decision-changes leave nothing intact: no absolutely invariant structure, no inequality that would be strictly repeated. They modify or affect all possible layers of philosophy. Even what we called the philosophical semifractality, the empirico-transcendental fraction, or the value of its dimension, tends to be effaced or is at least constantly scrambled. Even the "philosophical decision" can be identified as the structure of a philosophy only at a certain degree of stable resolution. It can no longer be identified when we move to another philosophy; at that moment it requires a supplementary interpretation in order to be identified. In a regime of sufficient philosophy, any philosophy is and must be interpreted by another, whether this other is explicit and elaborated or whether it is a question of implicit and poorly elucidated philosophical "positions." And yet this procedure of auto-/hetero-interpretation erases not only the singularity that philosophers speak of but also the singularities that philosophies themselves are. The interpretation of one decision by another, thus by Tradition—every philosophical decision represents the Whole of the history of philosophy for another decision—produces the global effect of inserting decisions into a *reserve* for which they work and that draws a surplus value from this labor of differences.

This is why the concept of "philosophical decision" and its descriptive value have to be relativized: they have a very limited scientific or fractal value, as we said; they are only pertinent from the perspective of another particular philosophy and according to the degree of resolution it represents with respect to and "on" the others. This weakness, this limited consistency of philosophical semifractality, its lack of *reality*—which makes it dissolve in scale-and-decision variations—obviously means that the auto-/hetero-interpretation, the reciprocal reading of philosophical systems, is caught within the circle of a Tradition that divests it of virtually any scientific value. "Decision" and the affiliated concepts ("multiplicities," "differences," "différance," "language games," etc.) have no theoretical stability or invariance; they are dragged along in a general circularity: of the structure and the detail, of the scale and the structure, of the tradition and a decision, of one decision and another, and so forth. Furthermore, the idea of *scale-and-decision variation* has to be, as we saw, critiqued and differentiated—we will say "dualyzed"—from the standpoint of science.

As we just appealed to it, this idea still expresses a philosophical decision or position. It cannot extract genuine invariants, a stable fractality with a theoretical value, insofar as it is affected in some sense by itself, insofar as it *modifies its concepts or its essence as it is exercised*. Put differently—this is a confirmation of what we already know—philosophy's auto-/hetero-reflection is stripped of every scientific value and cannot found any theory of philosophy, any *mathesis of thought*. It is a semitheory, a practice that has theoretical aspects, but always accompanies theory with its limitation or its division.

FROM PHILOSOPHICAL MULTIPLICITIES TO THE FRACTALITY OF PHILOSOPHY: FROM DECONSTRUCTIONS TO ARTIFICIAL PHILOSOPHY

Scientific knowledges' fundamental complication and chaos and, in this way, those of philosophy escape every philosophical enterprise, which only thinks of totalizing, systematizing, and capitalizing knowledges in an encyclopedia instead of knowing them and proposing a theory of their multiplicity. How can we emancipate the "labor force" at work in singularities, language games, disseminations . . . from this Tradition, from Philosophy itself [*LA* philosophie]? Only a science can conserve them without capitalizing them; it safeguards them in their *identity*, i.e., in their fractal constancy. It is obvious—this is a rule of every possible theory; philosophy is opposed to it—that thought must seek the most radical reason for the greatest fractality, that which can serve as a measure for philosophy itself, instead of letting itself be swept away by philosophy's objectives and its precipitation. Only a science can propose *the rigorous theory of a real fractality or chaos* and impose it on the philosophical decision without reconstituting the illusory project of a metaphilosophy. A theory in the scientific sense is never a metaphilosophy; it is only the means by which the real determines knowledge in-the-last-instance. It alone can salvage the concept of fractal dimension from its subsumption by the philosophies of singularities, multiplicities, catastrophes, and so on. And *it can thus let be the chaos of sciences without philosophically reinterpreting it once again, as "anarchism," for example.*

Using GF, a science of philosophy will not of course struggle directly against this clutter, this hesitation, or this wavering, against these balanced amphibologies. It is not a matter of returning to a geometry or a topology of philosophy; of proceeding through repeated approximations and idealizations, through the outline of more regular curves, through procedures of regularization and smoothing, which could be gradually substituted for the doctrines' poorly controllable details and would not perfectly track their contour. Such increasingly simple curves belong to the philosophical decision, and the history of philosophy itself does not fail to draw and trace these long chains that, without necessarily having a reason, form a tradition all the same. It is also not a matter of cutting or of differentiating these lines in an increasingly tight way, of crisscrossing these chains, of interpolating a manifold of decisions, of approaching the philosophical complexity of details: this procedure too belongs to philosophy; philosophy exploits and can describe it. Philosophy itself has already realized the "passage to the limit" (cf. the idea of an *unlimited becoming-philosophy*), either in the infinitely small or in the infinitely large of its objects. Regularization and uniformization, standardization and idealization, are procedures of philosophical *possibilization*, means of its *realization*; they are not at all *real* ingredients of this philosophy.

From the philosophical perspective itself, or from the perspective of autointerpretation, there is in general what may be called a *variation of the manifold*, of the *content in approximate diversity* of a philosophy. It is not so much this content that we are aiming for; or, if it is, then we are aiming for it from the perspective in which it is the material of a fractal constant of philosophy in such a way that this property is represented by an independent structure of the chosen method of approximation. The problem concerns the *dimension of the real or fractal content* (cf. the introduction) "of" philosophy. This content cannot be reduced to "multiplicity" or "partiality" as intraphilosophical concepts; to the thematic or other (signifying, textual, informational, etc.) content of systems. It cannot be appropriated and is not negotiable in the (hermeneutic, dialectical, analytical, etc.) autointerpretation that philosophy spontaneously gives of it. The problem is in fact double: to identify, if not to measure, the *reality* (not possibility) *content* and the *theory content* of a philosophy and of philosophy itself; an identification that is immediately the fractal practice of synthetic statements.

GF's theoretical importance lies in its displacement of the weakened or continuous forms of identity, but also of the différe(a)nces or singularities that form a system with them; at the same time, it leaves intact the Identity-of-the-last-instance, the reality it obtains from it, and this reality prevents it from dissolving in philosophy. Its importance vis-à-vis the philosophical themes of irregularity, interruption, or else catastrophe resides in its capacity to afford them a radical indelibility—their nonconservative conservation—against every "founding generality," every "full body," every "tradition," and so on.

It is well-known that the most innovative contemporary philosophies, those that defined a new postrationalist and postmetaphysical epoch of thought, are founded on the primacy of the Other instead of Being, a primacy exerted through numerous variations that do not concern us here. These philosophies have the following in common with their predecessors: they move within *the forgetting of the One qua One*, within the refusal to think Identity qua Identity. The more the *Identity-of-the-Other is forgotten*, the more philosophers overbid on this Other. They confuse this identity, which never *strays* from itself, with what constrains or enslaves the Other. They do not see that the Other's ultratranscendent position is just as arbitrary as any ontological decision and that it avails itself of metaphysics' old and most fundamental procedures. *By contrast, GF is the discovery of the Identity-of-the-last-instance of the Other itself. It is a theoretical discovery rather than a decision or an affirmation*: an immanent discovery, which concerns the essence of knowing rather than an object that can be located in the World. It is thus possible, in accordance with science's procedures, to use this new theoretical tool for other fields of knowledge, to require it as a theoretical criterion for philosophy itself, in order to produce knowledges that will fractalize philosophical continuity. We are obviously led to treat philosophy—in its specific universality—as a new class of theoretical objects, at least a new class of materials.

Accordingly, if A Phi must have some points in common with deconstructions, as GF does with restrained fractality, it is a matter of an appearance—an objective appearance, of course, but to which we will not give any reality. A Phi is a scientific and positive project whose sole program is philosophical innovation rather than anamnesis. It contains a critique of philosophy only in the second and not the first state, as is the case with deconstructions. GF was not the philosophical transposition and interiorization of geometry, any more than the non-philosophical synthesis of statements can be the philosophical extension of the synthesis of images or the extension of the computer's informatic possibilities. Here we will elaborate theoretical tools and research programs rather than new "positions." "Non-philosophy" is such a program, and GF is what allows its realization by rendering an authentic A Phi possible. A Phi does not result from a philosophical extension of AI—we ruled out this solution; it is a relatively autonomous concept that ultimately should be written as "artificial non-philosophy." But this would be a pleonasm, and, for reasons of strategy and of economy, it is better to speak of A Phi.

GF'S REAL ALGORITHMS

With GF, we have, in some sense, described knowledge's essential a prioris. The next and final problem is to "realize" or fix the conditions of the "empirical" use of this fractal a priori in philosophy. As we saw, the most "natural" or immanent application of generalized fractals concerns the artificial synthesis of statements, which at once simulate philosophical decisions and dogmas and absolutely exceed their norms of admissibility.

We already described the conditions of a de jure "application" of GF to natural language, which is grasped not in an arbitrary use but within the philosophical regime. More than an application: a determination, as non-philosophical or artificial, of statements whose only material is philosophical. They are the rules that stem from science's aprioric dimensions (cf. part 1), and these a priori rules are the rules of fractality: GF is in effect an a priori structure.

From this point of view, what we called elsewhere (in *Philosophy and Non-philosophy*) the rules of non-philosophical practices are obviously nothing more than fractal (albeit a priori) algorithms; they implement GF and can serve to describe philosophy's reality content. As obviously nonmathematical, real algorithms that use natural language, they allow the "non-philosophical" description of the structure of scientific knowledge and the recourse to philosophical statements.

Why "real" and not logico-mathematical? A computer synthesizes images on the basis of a description, of a prior fractal modeling—one that can be dominated by algorithms. We should thus distinguish the operation's theoretical aspect and its technological aspect. On the other hand, theory in the full sense we give to it (for example: *non-philosophy* in the case of first science) can no longer draw this distinction since it is "Identity" in-the-last-instance. It is at once, in an undivided way, the theoretical fractal modeling of philosophy and its "synthetic" image, at once the scientific explanation and the production of an "image" of philosophy whose internal structure cannot be discovered within philosophy. It is "synthetic" par excellence in the sense that its last reason is strict Identity and in the sense that it absolutely exceeds the philosophical givens or materials with which it is produced (it is therefore vis-à-vis philosophy alone and from its perspective that it is characterized in this way).

TRANSCENDENTAL DEDUCTION OF THE FRACTAL A PRIORI: FRACTAL DETERMINATION OF PHILOSOPHY AND PHILOSOPHICAL OVERDETERMINATION OF FRACTALITY

What is GF's field of validity? It is no longer the geometric and physical regions, the regions of so-called natural objects, but *thought itself in its most universal form*. First, science, as representation or knowledge; second, the "application" of science to philosophy, i.e., "non-philosophy." GF is a theory that stems from a particular science, the "first" science of Identities, but it holds for the essence of sciences, then for philosophy.

GF directly addresses the philosophical continuum, which enjoys a powerful universality; it shows in this way the simplicity, the diversity, or the scope of its "applications."

To be sure, in GF's relation to philosophy, which serves as its support and material, it is not a question of an *application*. A science does not *apply* to its object: it is a genuine *a priori determination of this material as object*, in this case as fractal object (i.e., object of knowledge).

We elucidated in the foregoing, though hastily, the reasons why GF was an a priori susceptible to a genuine *transcendental deduction* on the basis of the *dual*, the "relation" of Identity-of-the-last-instance to the philosophical material. The causality of the last-instance means precisely that the cause—the immanent real, the transcendental instance—cannot act directly on the *data* of experience, that it must necessarily pass through an a priori structure, in this case GF, which is *de jure fractality-for . . . the philosophical* or animated by a nonobjectivating intentionality (to) the philosophical. This transcendental deduction is obviously the key or the solution to the theoretical problem of A Phi.

The reason is that GF is the sole immanent fractality; it is more than a property: the immanent criterion of thought. All the others (geometric or philosophical) are transcendent and legible in the figure of the World. GF is immanent by its "transcendental" cause; it is located in itself rather than in the World and its geometry. It does not have a "home" within philosophy, but is exercised and manifested *on the occasion* of the World and its geometry and for them.

This a priori constant thus obtains from the Identity-of-the-last-instance its nonmathematical reality, which makes it touch on the *essence* of knowledge instead of being an ontic and regional knowledge. This does not mean that it is not a knowledge, but it belongs to first science and bears on science's *essence*. Its cause thus relates it immanentally to philosophy and gives philosophy the fractal dimension it lacked. But it does not give it this dimension in order to fill the gap and fulfill it as philosophy; rather, it is a "grace" that lifts it *outside* itself and *from* itself. A philosophical decision is never fractal on its own, nor does it *become* fractal from its own depth. It becomes so from the very cause of fractality, ultimately from fractality as cause. Everything that forms an edge, a border, an angle . . . within a philosophy becomes really fractal only outside itself: within the element of knowledge. Philosophy *is* not fractal; it is fractalized heteronomously.

Imposing on philosophy and its system a static and a dynamic—chaotic rather than simply differential or deconstructive—is first of all a question of right or of deductive rigor . . . Philosophy is GF's condition-of-object or of-material (it is thus philosophy and philosophy alone that is fractalized, and the other phenomena are fractalized as a result or inasmuch as they are virtually philosophizable). Other reasons are possible, other benefits can motivate the enterprise; but they are consequences of this deduction whose sense or "possibility" remains transcendental. For instance: the chaos languages are, by definition, extremely plastic and free as much as an axiomatic can be; they tolerate unstable environments and unforeseeable events—like the arbitrariness of the philosophical decision, since generalized chaos is (non-objectivating) intentionality (of) philosophy. These reasons can become goals; they remain secondary and cannot legitimate A Phi.

A methodical consequence that clarifies A Phi's meaning: instead of presupposing givens of natural phenomena, even of philosophy, in order to discover within them traces or regions of chaos and to expose them to a presumed and arbitrarily defined normality and dysfunctioning, we treat immanent generalized chaos as a guiding thread, no longer of a *reading* but of another *practice* of philosophy. It is the produced statements that matter as fractal rather than the elaborated statements. GF thus makes it possible to "describe" the philosophical phenomena nonspecularily, in a non-philosophical and noncircular way. To break at once with their "interpretation," their "functioning," and their "deconstruction," procedures that collectively respond to a supplementary decision, which dooms the game to a zero sum. This clearly does not amount to "superimposing" on or "applying" to them a "grid" that is allegedly other than philosophy. On the contrary, it amounts to transforming philosophy into a simple overdetermination or mode of existence of the fractal structure.

We will say of these synthetic statements of non-philosophy that they have, as their structure, science or fractality, which is *determinant* of their production and as their existence or conditions of existence, the philosophical, which is strictly *overdeterminant*. They are now philosophical only in their appearance or their occasion and through the channel of their material alone, which is also what overdetermines them.

GF is an a priori that is essentially indifferent to philosophical decisions or differences, which it nevertheless determines. They now play only an occasional role as materials or conditions of existence for this non-philosophical fractality. They no longer coconstitute it, and it can determine them as a fractal manifold endowed with a stability, an objectivity, and, above all else, an identity that they do not immerse in an interminable becoming. The very notion of "object," as we saw, loses in this way its vulgar and perceived sense, which extends into the philosophical idea of objectivation. But also the vague intuitive sense it has within the concept of the fractal geometric object. It is a matter of an objectivity that is immanent by its essence, manifest through and through and with a more than "objectivating" or "intentional" fractal structure or again intentional, but absolutely not positing. The scale-variation is relativized and transformed into an occasional condition in general, as we already explained, a condition that is necessary only for the philosophical existence of fractality, but is contingent for the *essence* of fractality. With this shift in status, it ceases to affect fractality itself, while allowing it to "be realized." Or, rather, it allows it to exist no longer under purely "philosophical" conditions, but under conditions that have a *philosophical origin*, and to exist as fractality *for* philosophy. Thus, as we have already repeated, GF is the theoretical instrument that should allow us to concretely implement the *science of philosophy* we are looking for.

FRACTAL MODELING OF PHILOSOPHY

In first science the transcendental Deduction of the a priori is also the right of philosophy's fractal modeling. The fractal a priori determination of philosophy is also what we were able to call—always under the aforementioned reservations or transformations—not only a scientific operation of modeling, but a modeling of philosophy by science as such. The philosophical objects, accompanied by their mixture-form, from which they are inseparable, can be *modeled* by *the essence of science itself.* They are not idealized and interiorized or else deconstructed, but "reduced" into "materials" for knowledges. Science does not represent an abstract fractality, but a fractality *for* them or a fractality with a transcendental origin.

GF therefore presupposes a transcendental (but not philosophical) concept of the *model* as *model-for* . . . (an object to be known)—a modeling intentionality, if you want—and thus presupposes that it is not an ontic knowledge, but the very essence of science, that can be this model-for . . . philosophy.

Modeling, as we will incessantly see, is not a specular double, a simple image that reproduces a given object in a "theoretical" mode. It is a *knowledge* that, in a sense, is without "common measure," without a third and transcendent or "common form" with its object. It is an absolute or absolutely mirrorless reflection. It proceeds through phenomenalization in the theoretical mode; it manifests the philosophical and, in a sense, the philosophical *itself*—but, from now on, in theoretical forms. This manifestation in the mode of knowledge takes the form of a phenomenal state-of-affairs that shows itself absolutely (theory is a nonpositional reflection (of) self . . .). Thus this reproduction, the very reproduction of knowledge, does not reproduce philosophy in a specular double; it reproduces it, but without its fractalization, which gives to it a scientific universality. This is, moreover, why GF is a *model* rather than an interpretation. Whereas the restrained, geometric, fractal modeling ceaselessly risks turning in an epistemological circle of the *object* and of its *mathematical image*, as a representation that reproduces an object and risks ceasing to be a model, GF is what "separates" science from philosophy, that through which science separates itself from philosophy, discharges it while summoning it, and dissolves the amphibologies.

It is, moreover, not a matter of a geometric, philosophical, or mixed *idealization*. The theory contains an ideal dimension, but an ideality devoid of every transcendent object it would have reproduced. GF manifests precisely philosophy's transcendent *figure*. From this point of view, it deidealizes philosophy at the same time that it dematerializes and definalizes it and, having suspended every teleology, delivers it to chaos.

Given this function and this essence of the model, we will only say that GF models irregularities *of* philosophy, situations of fragmentation, or the bristling, the angularity peculiar to the details of the philosophical Tradition as a curve of metaphysical differences or decisions. For philosophy is enough to interpret such situations; and they define its essence. GF models philosophy only by producing knowledge from it, but without any originary continuity with it. We can see here how much the idea that "science studies an object's properties" is problematic and poorly formulated. It is a matter of a property of scientific knowledge, which is thus in its turn a knowledge produced by and in the theory of science, in first science.

PHILOSOPHY'S CHAOTICIZATION; PHILOSOPHICAL SYSTEM AND SCIENTIFIC CHAOS; NON-PHILOSOPHY'S CHAOS-LANGUAGES

We typically examine philosophy with procedures of magnification or of enlargement, with scale-and-decision variations. But these procedures become fertile and are distinguished from the Greek and average gaze with which philosophy tends to be confused only when they are placed in the service of a fractal experience that, in itself, does not depend on them. The science of philosophy is not a variation in "resolution" or even in decision, which are only an occasional condition of fractalization and not its essence. In principle, we see the philosophical under non-philosophical conditions and "in-fractality"; we describe the images that fill this fractal vision and that are detached—by transforming themselves—from philosophy's inert body. These images are more irregular, more disordered, than the philosophical clutter or fuzziness. But this is a difference in nature and not in degree. At the very scale of philosophy, of its autoreflection, all that can oscillate from the most continuous to the most dispersed, from the most regular to the most aleatory, is grasped at the heart of theory as a fractal chaos of non-philosophical knowledges. It is more than a magnification of details and a variation in the optic field; it is a mutation in the very conditions of thought's "optics."

With the science of the One, we can cast on philosophy another gaze than History's, Tradition's, Sending's, or Destiny's. We can see philosophy in-Identity, that is to say, "in-"fractality and no longer as a more or less continuous curve. Non-philosophy no longer consists in drawing lines, but in fractalizing the lines, even when they are already affected by folds or points of retrogression and dispersion, by critical points or catastrophic breaks. Philosophy is not itself (from its own viewpoint) fractal, but we see it in this way within the element of another discourse that makes use of it without obeying it any longer. For theory is "vision-in-science" and is filled with philosophical materials only by imposing on them a new "chaotic" distribution. This distribution does not lack rules, as we suggested, but is the fractalization of philosophical rules by rules of a whole other type. There is an a priori intuition (of) fractality; it "surpasses" the "transcendental imagination" (philosophy's resource), not metaphysically but "chaotically." By means of knowledge, it is capable of receiving not only the structure of lacunary clusters but more profoundly the structure of chaos that orders these clusters themselves.

Here and there (Plato, Descartes, Kant, Husserl, and Nietzsche), philosophy has sufficiently brandished the menace of chaos—in order to propagate the rules of the philosophical Decision it imposes through this violence—that one should finally take it seriously, respond to the challenge and put an end to the blackmail. There is really a chaos, but it is the chaos of the real and of the knowing of the real, which philosophy has never mastered. Philosophy simply called on this chaos as a scarecrow, which, when it is exposed and manifested according to its own laws, is capable of subsuming philosophy itself. We limit philosophy as a simple disorder in view of a transcendent order, as a fear of chaos so as to make way for science as the sole theoretically valid chaos. This chaos is founded on the minimal order of Unilaterality and is capable of containing the philosophical Decision's rules or half-rules.

Any philosophy whatever is a system and not a chaos; it is therefore a circular structure. But more profoundly and in the full phenomenal sense of the term, it is a "linear" system (with lines, vectors, or directions, with continuous trajectories and curves). Generalized chaos is a static and a dynamic that are delinearized, absolutely *fractal through and through*; fractality is no longer sustained by the support or the object, but by Identity-of-the-last-instance. It is not, in fact, a marginal or an atypical angularity, which can be partially compensated for and negotiated, rectified and idealized: ideality and objectivity themselves are intrinsically fractal; they are as "multiple" or individuated as the Identities-of-the-last-instance. Hence A Phi ceases to trace lines and to recognize the chains of a Tradition, to identify the signs of a consignment and the critical points of a process. It describes ever new fractal dispersions, radically diachronic, in which it is chaos as the originary space-time of fractality that occurs, forever outside every teleology and philosophical closure that should be passed through, circumvented, leapt over, etc.

Non-philosophy multiplies—through the play of realizations of the simplest fractal structures—more than an unfolded surface, more than the flattening and effacement of a fold that reverts to a surface: a structure that is, if we may phrase it this way, the very intermediary of all the possible differences, a unilaterality more powerful than every in-between. At the same time that it voids this dimension of every object that is supposed to be given in the World or in philosophy, at the same time that it absolutely unlimits it, non-philosophy fills it with local yet fractally dispersed clusters and fills it without saturating it: these are the chaos-languages. Non-philosophy is a practice of *chaos-languages*. Its material is *philosophy*-language (or language-in-philosophy); its theoretical means, generalized fractality; its product, language-in-philosophy, distributed in the mode of chaos. This distribution is more than a *dis-persion*, which remains a unitary concept close to difference. It is the dimension of an opening-without-open, of a space-time that occurs without being able to fold or close on itself.

It then becomes possible, with very little philosophy, with a very poor material, to produce an extreme variety and an extreme complexity of statements that make one last reference to philosophy, call it up one last time without being able to localize it. GF makes it possible to stretch out, infinitely and absolutely, outside every pregiven surface or horizon, philosophy's little stain, to stretch it out according to an ongoing, absolutely spatiotemporal dimension, but not according to a line or toward a horizon. Chaos neither is nor becomes [*devient*]; it absolutely occurs [*ad-vient*] each-time-one-time. The horizon (goal, telos, end) is merely the material of fractal chaos. What we previously called philofiction or hyperspeculation is none other than this fractal practice of language-in-philosophy or of virtually philosophizable language.

The non-philosophical statements produced in this way enjoy a special property that the classical philosophical statements, as *differentiated* as they may be, lack. The statements fabricated more or less artificially with procedures that remain *in a dominant way* those of the philosophical continuum (the statements of deconstruction or of language games) also do not enjoy this property. On the contrary, every change in the degree of resolution on these statements or in the decision that serves as their material, every new production of statements, allows an inequality or unilaterality to subsist. This inequality does not cease to occur, under variable conditions of existence, but it is lived and received not as identical *through* these variations, but as the Identity (of) their unilaterality or their strangeness. New details are uncovered. More exactly: new statements are produced or occur (the equivalent of the interpolation to infinity). But they have this radical fractality they cannot eradicate, which is inalienable in its conditions of existence. The unforeseeable is here more than aleatory; it is lived "in itself," as fractality in the flesh rather than as a simple fluctuating and vanishing property in its material.

These are particularly nonintuitive statements; they are foreign and unacceptable to every philosophical logic. And yet they are deployed in their own dimension, the dimension of an unreflected and opaque intelligibility. GF is said of statements whose form, *the form of their sense*, is irreducible to any philosophical syntax. These statements are almost meaningless according to philosophy and pure grammar (cf. Husserl), and yet they are not contradictory; they are produced with coherence and rigor according to the criteria of a real intelligibility (even if it is philosophically blind). Philosophical but not scientific "monsters," they are useful for opening to theory new, already existing fields (Technology, Ethics, Aesthetics, etc.) that up to now have been subject to and exploited by philosophy alone. They can be directly used for deciphering the reality of science, technology, and philosophy; for rendering them intuitive in another mode; and for rendering Transcendence accessible in a mode that is no longer itself given as transcendent. Philosophical intuitivity (so dependent on natural, perceptive, and topological intuitivity for example) is replaced by another that is less external or contingent: a pure theoretical intuitivity, which is the element of this real and no longer logical, fractal axiomatization.

THE REALITY-EFFECT OR SEMBLANCE-WITHOUT-IMAGE OF GF; SPECULAR REALISM AND REALISM OF-THE-LAST-INSTANCE; SCIENTIFIC MODEL AND PHILOSOPHICAL MODEL

A realist appearance of a special type accompanies GF; it has to be described and should not be confused with the fully specular realism of an *object*-image.

There is a very intimate connection between the self-similarity of geometric fractals and their power to simulate the real. The connection is even more intimate between GF's Identity-of-the-last-instance and the realist appearance it creates. A few clarifications have to be introduced with respect to scientific representation, to its power of *manifestation* of the real, of its cause-of-the-last-instance. As we understand it, science, on one hand, does not manifest its *data* or the theorico-experimental phenomena that serve as its material; it manifests the real—its cause—by means ("occasion") of these *data* and the objects of knowledge of which they are the ingredients. It knows the real by means of the production of objects-of-knowledge; it does not know these objects. And, on the other hand, it is a *nonpositional* reflection, a description, *(of) the real*, a theoretical representation inasmuch as, in itself and by its global or undivided existence, it is the manifestation-in-theoretical-mode (of) the real as it is, i.e., inalienable precisely in its representation. The non-philosophical or artificial statements created by means of GF do not represent an "object," the material, in a specular mode; they are the nonspecular, albeit purely descriptive, representation of Identity-of-the-last-instance, their cause. This is an absolute manifestation, without reference or aim; it manifests the real of-the-final-instance as it is, although only by its existence as reflection and not by an image that would biunivocally correspond to an object or that would represent in the manner of a *tableau*. There is indeed *a reality-effect* given by scientific representation, but it is intrinsic, absolute in its order, not locally motivated by reference to a transcendent object. *It is a reality affect, an invincible and nonlocalizable, nonidentifiable feeling of reality—what we should call a realism of-the-last-instance. In-the-last-instance alone, and despite everything, despite the absence of every relation of resemblance or of "tableau" to an object, science presents itself globally as an index of reality, of undivided "realist" representation.*

To better understand this apparent paradox, let's start again from philosophy. It carries an objective realist appearance, which is necessarily divided or separated, half-realist, half-imaginary. It gives itself in the mode of Tradition, of Metaphysics [*LA métaphysique*] and of its History, of a Memory and a Reserve—of a philosophical *Great Body* that insists and subsists, watches over the work of philosophers and necessarily accompanies them. Philosophy [*LA philosophie*] is a half-invariant, half-variant process. The identity according to which it is reproduced is not given as simple and inalienable, as identity-of-the-last-instance; it is alienated and returns to itself in the form of a Same, a Great Appearance—an "objective" appearance, i.e., binding, but divided, lived as partially real and partially ideal. The philosophical landscape is at once stable and mobile—metastable; at once "real" and an "image"—surreal; modified by every new decision or every scale-variation. Hence its uncertainties and its objectively doubtful or problematic nature: philosophy is an *insufficient Identity*; its nature as lack-of-the-real [*manque-au-réel*], its visceral nihilism, its limited consistency, plagued by the simulacrum; its ultimate loathing of every "realism"; the extensive impregnating of its thought with artifact and imagination; its tissue of transcendent and intuitive images or metaphors (circle, whirlpool, band, ground, horizon, cone, ribbon, surface, plateau, etc.).

Science's realism has a whole other nature. In a sense, it is *without division*: one can only say that science as a real appearance or a *semblance-without-image* is *equally* imaginary by one of its sides. In fact, it does not have two sides; it is uniface or unilateral. Undivided through and through, it is an integral (re)semblance (to) the real. But (to) a real of-the-last-instance alone: its realism is an undivided effect; it is not founded in the reproduction or even in the simulation of a transcendent object that would divide or nuance its reality-effect. Science, thanks to its fractal nature, emancipates thought from every ideal of specular resemblance—which is sometimes more or less *deferred*—to supposedly given objects and to their horizon: the World, History, Power, Philosophy. The remainderless destruction of the philosophical theory of "truth tableau" or of "common form," of transcendent mimesis, is included de jure in the scientific practice and in the nonpositional-Reflection (of) the real.

If scientific realism is an appearance that is real as appearance, in its order of representation, if there is a relative autonomy of the objective appearance of reality that science is, we should say that there is a fractal realism of the same nature, that GF communicates an undivided or in-the-last-instance reality-effect to statements whose synthesis it enables. This originary impression of reality (strictly without foundation, but not without cause) affects all knowledges produced under the "fractal" rules of science, but outside the codes that ground the resemblance to philosophy. There is indeed a self-resemblance or a realism proper to this philosophical material, but it remains secondary, included in an undivided impression of resemblance (—to . . .)—to nothing transcendent. An undivided semblance that is not founded on any reference, any object-image. It is a reflection-without-mirror or a (re-)semblance-without-image, "abstract" and a priori; but it completely penetrates all the artificial statements. The philosophical variations in scale and in decision, in material, do not cut into it: it is reproduced, or rather occurs, as fractality itself. It is nothing other than the *objective appearance of the fractal structure*, how it is manifested. *And it is manifested as real without nevertheless reproducing for its part transcendent images of its material, without folding back on the fractal structure itself.*

This fractal realism does not bear on the used and fractalized material, for it is not this material that is fractally reproduced. It bears on the fractal structure itself. It is, moreover, clear that there are phenomena of autosimulation peculiar to philosophy. But the fractal reality-effect does not prolong them; it does not therefore oscillate between transcendent and imaginary reality, between copy and model, between one simulacrum and another. The philosophical "family resemblance" of non-philosophy is one thing (it moreover varies according to the type of assumed decision); the fractal production of non-philosophy as representation (of) the real is another. It is now impossible to blend them unitarily. Science clarifies the uncertain and unstable realism of philosophy; it destroys philosophy's amphibological nature (half-real, half-imaginary) and restores to the order of fractal appearances (of) the real its *identity*.

A Phi's theoretical context, as we see, excludes the model/copy system ("model" in the Platonic-metaphysical sense, not in the scientific sense), but it equally excludes its inversion and displacement—its "Nietzschean" becoming—in the unlimited simulacrum, in the becoming-simulacrum. Just as GF is distinguished from philosophical semifractality, so the fractal modeling *of* or *for* philosophy is distinguished a priori from philosophical concepts of models and thus from simulacrum and simulation, since it fractalizes them in their turn. Identity-of-the-last-instance is the cause that maintains the order of the model, of modeling, as relatively autonomous. It does not blend once more with this order so as to generate the famous "simulacrum" of postmodern philosophies. This simulacrum results from the autodissolution of the circular "model/copy" system, of the *decision* of the model and the copy, instead of receiving its reality from an unbreachable Identity. Of course this dualysis, this uni-lateralization of the unitary and undetermined concepts of resemblance, realism, reflection, and model, is equivalent to their fractalization. Not only the reality-effect is essentially fractal, but its description also has this character.

GF's rules or algorithms allow us to synthesize statements and texts that cannot be reduced to the codes of philosophical perception and intelligence. But these statements do not objectively appear to have any less of a striking reality (and validity), opening a prodigious non-philosophical heaven, an intelligible universe superior to every psychological and philosophical imaginary. In the domain of thought, first science—in the extension of effective sciences—produces an effect comparable to that of the artificial synthesis of images in the visual domain and in cinema, for example. It creates an "artificial" space and an "artificial" time, a nonontological opening that traverses the "Being" of philosophy, intellectual-and-sensible landscapes, simulations of scientific and philosophical behaviors, and so on. The equivalent of a visual and graphic informatics would, in this case, be an "automatic" and non-philosophical writing of philosophy, an entire scientific conception assisted by philosophy—at least by artificial philosophy or non-philosophy.

These "quasi"-philosophical, synthetic, absolutely universal space-times also produce exploitable aesthetic effects or poetry-fiction.

First science thus opens a new "space-time" for the exercise of thought, unchained from the World or History, but not imaginary for all that. It opens a sphere of existence that can be called intelligible. Realism is inevitable; no thought can escape it, and it would be nonsense or absurdity to want to dispense with it, an effect of philosophy's perverse realism. The problem is to liberate realism from the constraints of supposedly first Transcendence, from the ontico-ontological disorder. This disorder is not chaos, and its function is in fact to *normalize* science's chaos and to bring it back into the philosophical order. Far from being imaginary or semi-imaginary, science is the real critique of philosophical appearances and effects of reality, the critique of the more or less idealized trompe l'oeil realism that is desired by all the philosophical decisions without exception. But this critique is an effect; it is not one of the objectives of first science.

"NATURAL" OR SPONTANEOUS PHILOSOPHY AND ARTIFICIAL PHILOSOPHY; SCIENCE OF SENSE; OF NON-PHILOSOPHY AS MONOGRAPHIA OF THE REAL AND OF THOUGHT

From the start—it is even there that they were invented—fractals were tied to problems of economy, meteorology, and linguistics, to natural phenomena that, as always, seemed to go beyond their then possible mathematical modeling. Once this step is taken and this law is summoned, it is up to us to note that science itself is, in a sense, a "natural" phenomenon that goes beyond the "modeling" (or rather the interpretation) carried out nowadays by philosophy and epistemology. We can also consider philosophy to some extent as a quasi-"natural" process or object. There is a *spontaneous*, in some sense ambient *philosophy*, a *philosophical faith* whose spontaneity feeds the elaborated system; it is never put into question by philosophers themselves, who do not cease confessing this faith. Despite their statements to the contrary (in order to mark the exceptional character of their discipline), philosophy's existence is a mundane phenomenon, as frequent, as inscribed in the World as the topology of natural or linguistic geometric forms. This ambient philosophy can be autodescriptive, it can fold on itself or prolong itself more explicitly. But it cannot describe itself in a theoretically rigorous and grounded way; it cannot produce validated, rectifiable statements under conditions of scientific stability, and still less can it provide any explanation of itself.

This is where first science and its fractal tool intervene: to scientifically describe this ambient and natural, quasi-vital philosophy, to provide an essentially theoretical description of this set of invariant gestures (the "philosophical decision"). It is possible only if, far from unmarking or untracing the philosophical in a metaphilosophy, the description takes a detour through the generalized fractal and is practiced as a modeling of philosophy.

Philosophy will have only delivered thought from its interior and most representative images—and only for the benefit of movement, becoming, and the Other. It will have crammed it with more idealized, more interiorized images, but these images presuppose some continuity with perception, with common sense, with the topology of natural forms, and with the *intuition* of the World. The paradox is that it is science and not philosophy that has at its disposal the least transcendent realism, the least imaginary reality-effect, the effect that is best grounded in the most irrefutable real. Thus science alone can reform the philosophical understanding and lead it to a freer form of operation, a form that is less entangled in the insistence of the World's objects. *Thought of synthesis* (in the non-philosophical sense of the word, not in its informatic and cognitivist sense), which is also a *unified* and not unitary thought of science and philosophy, is the thought that receives its true "spirituality," its dimension of free transcendence in relation to the philosophical Tradition, the philosophical Reserve. Its nature as GF allows it to create new research dimensions: politics-fiction, poetry-fiction, simulation of thought's behavior, etc.

The solution to the problem of a plausible, philosophically and scientifically noncontradictory A Phi resides in the discovery of a fractality that concerns knowing rather than nature. It thus concerns the (philosophical) sense through knowing, even if it is devoid of sense and of any reflexive dimension. The solved problem is that of a *science of sense*, which is de jure impossible for philosophy itself as well as for the science that is used and understood in a positivist way. Obviously the problem for us is no longer to subsume new objects under philosophy, not even its old objects, which escape "mechanistic" science (catastrophes, differences, metaphors, and metastability of forms), to subsume them under a renewed regional-ontic science, subtler and more impregnated with philosophy (cf. the mathematical Theory of Catastrophes and its spontaneous neo-Aristotelian philosophy). Instead, the problem is to subsume, under science, philosophy itself, i.e., the object that resists it par excellence, more than a simple regional object that is still inaccessible to a current science. With philosophy, we are dealing with a supraregional, ontological, or "fundamental" object, and it is true that none of the existing ontic sciences, which are adapted to being, can assert its rights next to Being. We needed the discovery of a science of the One to discern the possibility of treating philosophy at last, and without contradiction, as the object of a theory, i.e., as a set of phenomena or *data* that would no longer arise from their autointerpretation and its phantasms, that would arise from a rigorous theorization. Paradoxically, it was necessary to show that science is unintelligible for the procedures of the philosophical continuum in order to uncover within it, in return, the power to fractally model the most elaborated phenomena of sense, the highest semantic and discursive layers—those of the philosophical use of language.

Artificial Philosophy is thus not a philosophy of artifice, the interiorization of foreign artifices into philosophy (aesthetic, informatic, etc.). Nor is it the contemporary philosophical implosion of simulation and of the simulacrum—it is a matter of breaking with this spontaneous, repetitious, and naively idealist philosophism. Nor, as we said, the simple transfer of criteria and practices from AI or cognition to *thought*. From our point of view, we have assumed the means of a conceptual and theoretical mutation capable of giving philosophy a new impetus—an impetus that does not come from its own depths, but from elsewhere. The principle or motor of this break in the theory of thought is the theoretical and unphilosophical discovery that the thought of the One qua One is, on one hand, of the order of a science and, on the other, is fractal or irreducible *to* every philosophical logic. This discovery, which was initiated here and there in some philosophies, did not receive its full sense until now. It cannot in fact leave philosophy's claims intact and has to contest its most global status within thought and the works of man. This contestation entails the renunciation—at least the de jure renunciation—of the narcissistic practice and procedures of the philosophical continuum; it entails the elaboration of a new nonepistemological theory of science and the thematizing of the fractalization of philosophy.

The majority of philosophies have therapeutic goals (therapeutic of the soul, the body, thought, language, logic, perception, etc.). But there is also a universal pathology of philosophy (fetishization, excess of memory and repetition, loss of variability, decline of differentiation, sedimentation, smoothing, or regularization, insofar as philosophy refuses to abandon their principle and identifies with their autoaffirmation). We find at once an arbitrariness of the decision, of the possible, and of the lack of reality, and what balances them (repetition) in the vain hope of providing, through this procedure, a lacking reality. We find not only the nihilist senescence: the internal decline of philosophy, but also congenital wisdom and old age: philosophy is born fatigued. Its fractalization by science, by the chaotic dynamic proper to theory, changes its practice and grants it the unexpected gift of a vigor that we may have believed to be lost.

What is non-philosophy? It is the *monographia of the real* and, thereby, of the World, as Identity-of-the-last-instance-of-the-Other. It is still a matter of Clouds, the Ocean, the Heaven and the Earth, but insofar as, far from being left to themselves, i.e., to the hubris of unbridled philosophy, to spontaneous autoposition, they are now described from the perspective of their cause, *from* this Identity-of-the-last-instance.

So it is the *monographia of thought*, its rigorous description, or its unified theory, that invalidates the hierarchical distribution of philosophy and of science. And a monographia should have good reasons, which have been theoretically tested (in the One itself), so that it does not appear as the return of metaphysics against the contemporary *digraphia*, the ever unitary digraphic decision . . .

NOTES

INTRODUCTION

1. The project of a "science of philosophy" has to be nuanced. It cannot mean that philosophy becomes the "object" a science *aims for*, but only that it is the *datum* or the *material* required by this science outside every "objectivation" and in view of a science of the *essence* of science, an essence that is its "real object." The concept of "object of a science" is ambiguous and will be dissolved. We distinguish the phenomenon-object or *datum* and the real-object of a science.
2. Obviously "artificial philosophy" does not designate the "philosophy of the artificial" or the "philosophy of Artificial Intelligence," but rather an "artificial" treatment of philosophy itself.

1. SCIENCE

1. A first, more reduced and less precise version of this chapter appeared in the journal *La Décision philosophique* (Editions Osiris, 1987). That text is expected to introduce a treatise devoted entirely to "first science" or science of the *essence* of science.

4. THE CONCEPTS OF GENERALIZED FRACTALITY AND CHAOS

1. *La Recherche*, no. 85, January 1978, p. 18.
2. Citations from *Les Objects fractals*.

INDEX